S198
Understanding Mars

Iain Gilmour

Mars as seen by the Hubble Space Telescope on 26 August 2003 when Mars and Earth were 55 757 930 km apart – their closest in almost 60 000 years.

S198 COURSE TEAM

Chair Iain Gilmour
Author Iain Gilmour
Course Consultant Colin Pillinger
Critical Readers Stuart Freake, Mabs Gilmour, Ian Wright
Course Managers Jennie Neve Bellamy, Isla McTaggart
Editor Pamela Wardell
Graphic Artist Sara Hack
Graphic Designer Jenny Nockles
Science Short Course Programme Director Elizabeth Whitelegg
Course Assessor Prof. Grenville Turner FRS, University of Manchester

The book made use of material originally produced for the S283 *Planetary Science and the Search for Life* Course Team by Philip Bland, Iain Gilmour, Neil McBride, Mark Sephton, Mike Widdowson and John Zarnecki.

Cover image: An artist's impression of the Mars Odyssey spacecraft in orbit above Mars. In addition to its own scientific objectives, Mars Odyssey has a crucial role to play in providing communications support for NASA's Mars Exploration Rovers and can provide similar support for Beagle 2.

This publication forms part of an Open University course S198 *Exploring Mars*. Details of this and other Open University courses can be obtained from the Course Information and Advice Centre, PO Box 724, The Open University, Milton Keynes MK7 6ZS, United Kingdom: tel. +44 (0)1908 653231, e-mail general-enquiries@open.ac.uk

Alternatively, you may visit the Open University website at http://www.open.ac.uk where you can learn more about the wide range of courses and packs offered at all levels by The Open University.

To purchase a selection of Open University course materials visit the webshop at www.ouw.co.uk, or contact Open University Worldwide, Michael Young Building, Walton Hall, Milton Keynes MK7 6AA, United Kingdom for a brochure. tel. +44 (0)1908 858785; fax +44 (0)1908 858787; e-mail ouwenq@open.ac.uk

The Open University
Walton Hall, Milton Keynes
MK7 6AA

First published 2003

Copyright © 2003 The Open University

All rights reserved. No part of this publication may be reproduced, stored in a retrieval system, transmitted or utilized in any form or by any means, electronic, mechanical, photocopying, recording or otherwise, without written permission from the publisher or a licence from the Copyright Licensing Agency Ltd. Details of such licences (for reprographic reproduction) may be obtained from the Copyright Licensing Agency Ltd of 90 Tottenham Court Road, London W1T 4LP.

Open University course materials may also be made available in electronic formats for use by students of the University. All rights, including copyright and related rights and database rights, in electronic course materials and their contents are owned by or licensed to The Open University, or otherwise used by The Open University as permitted by applicable law.

In using electronic course materials and their contents you agree that your use will be solely for the purposes of following an Open University course of study or otherwise as licensed by The Open University or its assigns.

Except as permitted above you undertake not to copy, store in any medium (including electronic storage or use in a website), distribute, transmit or re-transmit, broadcast, modify or show in public such electronic materials in whole or in part without the prior written consent of The Open University or in accordance with the Copyright, Designs and Patents Act 1988.

Edited, designed and typeset by The Open University.

Printed and bound in the United Kingdom by the Alden Group, Oxford.

ISBN 0 7492 6634 1

1.1

CONTENTS

INTRODUCTION 1

CHAPTER 1 SOLAR SYSTEM EXPLORATION 3
1.1 The Solar System 3
1.2 The planets of the outer Solar System 6
1.3 The planets of the inner Solar System 14
1.4 Summary of Chapter 1 23
1.5 Questions 24

CHAPTER 2 MISSIONS TO MARS 25
2.1 Introduction 25
2.2 The Mariner missions to Mars 26
2.3 The Viking missions 28
2.4 The Mars Pathfinder Mission 29
2.5 Mars Global Surveyor 30
2.6 Mars Odyssey 32
2.7 The 2003–2004 missions 34
2.8 Summary of Chapter 2 36
2.9 Question 36

CHAPTER 3 INSIDE MARS 37
3.1 Introduction 37
3.2 Some basic physical properties of the planets 40
3.3 How to make a planet 43
3.4 The internal structure of the Earth 50
3.5 The internal structure of Mars 52
3.6 Summary of Chapter 3 55
3.7 Questions 55

CHAPTER 4 THE MARTIAN LANDSCAPE 57
4.1 Planetary resurfacing 57
4.2 Volcanism on Mars 68
4.3 Impact craters 78
4.4 Martian fluvial and aeolian processes 94
4.5 Summary of Chapter 4 99
4.6 Questions 100

CHAPTER 5 MARS AND LIFE 103
5.1 What is life? 103
5.2 The martian environment 122
5.3 The search for life on Mars 136
5.4 Summary of Chapter 5 145
5.5 Questions 146

AFTERWORD 149

ANSWERS AND COMMENTS 153

GLOSSARY 160

ACKNOWLEDGEMENTS 165

INDEX 166

INTRODUCTION

Of all the planets in our Solar System other than Earth, Mars has been the main focus of discussions on the possible existence of past or present life on other worlds. Even in ancient times, both mythology and informed thinking suggested that Mars might be inhabited. This belief prevailed through the Middle Ages and even to modern scientific times. By the advent of the space age, the idea of any advanced life-forms had all but disappeared. However even in 1960, when the first space mission to Mars was attempted, the seasonal changes in colour on Mars that had been observed from the Earth were still taken by some to indicate that Mars was covered in vegetation.

Mars is named after the Roman god of war. In early Roman times he was god of agriculture prior to his military association.

Arguably the most notable of early scientific observers of Mars was the 19th century Italian astronomer Giovanni Schiaparelli (Figure 0.1) who made an everlasting contribution to our cultural awareness of Mars. In 1878, he announced that he had observed extensive straight lines or streaks on the surface of Mars (shown in blue in Figure 0.2). Schiaparelli treated his findings with great caution and at first doubted his own observations. But successive observations convinced him of their veracity. In his description of these observations, he used the Italian word 'canali' which means 'channels' or 'grooves'. This generated enormous interest and many scientists and writers in the English-speaking world translated this word as 'canals', implying artificial waterways presumably created by intelligent beings. For nearly a century the idea of life on Mars was embedded in popular imagination and was a fertile source for both serious literature (for example H. G. Wells' *War of the Worlds*, first published in 1898), and pulp fiction in its many manifestations (Figure 0.3). One of those most inspired by Schiaparelli's observations was Percival Lowell (1855–1916), who devoted his energy to the study of Mars (Figure 0.4) and proposed a theory that the 'canals' were a result of attempts by struggling martian inhabitants to irrigate the planet from the melting polar ice-caps!

Figure 0.1 Giovanni Virginio Schiaparelli (1835–1910), Italian astronomer and director of the Milan Observatory. He made meticulous observations of Mars (1877–90), including features that became known as 'martian canals' (Figure 0.2). He continued to observe Mars faithfully until his eyesight failed in 1892.

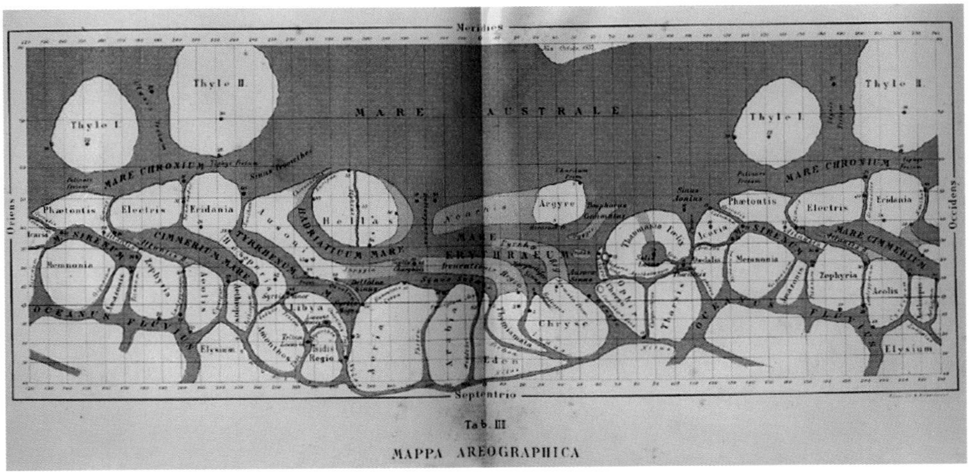

Figure 0.2 One of Schiaparelli's many sketches of Mars, this one completed in 1881, based on his telescopic observations. These led to a 'craze' amongst astronomers, both amateur and professional, to search for evidence of intelligent life and extravagant claims by some of positive evidence. Schiaparelli remained rather moderate in his assertions about 'canals', suggesting that they might be natural rather than artificial structures.

Figure 0.3 Illustrations from several examples of the treatment of life on Mars in literature. (a) The cover picture by an unknown artist of *La Guerre dans Mars,* (b) an illustration of a floating martian city by Paul Handy in *Letters from the Planets* (1890) and (c) a cigarette card from Wills's cigarettes.

Figure 0.4 Percival Lowell (1855–1916), a Boston-born American, who spent his life devoted to business, travel, literature and astronomy. He became widely known for his theory that the martian 'canals' were a result of attempts by the struggling inhabitants to irrigate the planet from the melting polar ice-caps. He founded the great observatory that bears his name at Flagstaff, Arizona, initially with the exclusive intention of confirming the presence of advanced life-forms on the planet. Although his theories met with widespread opposition, he received numerous honours during his life.

The dawn of the space age and the exploration of Mars and other planets by spacecraft brought an abrupt end to such speculative theories. In Chapter 1 we will look at what, in astronomical terms, is our own backyard, namely our Solar System, and at some of the space missions that, over the past 40 years, have radically changed our view of our small corner of our Milky Way Galaxy. In Chapter 2 we will look at the various space missions that have flown by, orbited or landed on Mars before going on in Chapter 3 to look at Mars in detail starting with what we know about the interior of the planet. Chapter 4 considers the martian landscape and the geological processes that have shaped it. Finally, in Chapter 5 we consider the key area that future space missions to Mars hope address: the subject of Mars and life.

CHAPTER 1
SOLAR SYSTEM EXPLORATION

1.1 The Solar System

For millennia, the stars visible in the night sky have been the subjects of study. Naturally, the Sun and Moon dominated thinking about the Earth's place in the Universe, but five planets (Mercury, Venus, Mars, Jupiter and Saturn) were well known to early civilizations because of their brightness and rapid motions relative to the 'fixed' stars. Indeed, quite sophisticated knowledge of planetary motions developed in several early cultures, defining the idea of a Solar System (as opposed to the fixed stars). However, in 1512 Polish astronomer Nicholas Copernicus, after studying records of naked-eye observations of the motions of the planets against the starry background, realized that the Earth went around the Sun. Furthermore, the German astronomer Johannes Kepler deduced in 1604 that the planets move in elliptical, rather than circular, orbits. It was not until the Italian scientist Galileo Galilei discovered four satellites of Jupiter in 1610, with one of the first telescopes ever made, that the existence of other bodies in the Solar System became known, and it was 1801 before the first asteroid (a 'swarm' of small rocky and metallic bodies between the orbits of Mars and Jupiter) was observed.

> Copernicus was not the first astronomer to suggest the Earth went round the Sun. The Greek philosopher Aristarchus had made the same proposal around 280 BC.

Since the invention of the telescope, the number of known bodies in the Solar System has increased enormously. Technological progress has led to occasional surges in that number. The construction of larger and larger telescopes during the 19th and early 20th centuries led to the discovery of many planetary satellites and huge numbers of asteroids. In the two decades between 1970 and 1990, spacecraft explorations of the planets revealed yet more satellites and entire planetary ring systems. Developments in the technology of light-sensitive semiconductors such as charge-coupled devices (CCDs), which form the basis of digital cameras, meant that photographic film for astronomical work was mostly abandoned by the 1990s. CCDs enabled astronomers to record virtually all the light passing through the telescope, allowing incredibly faint objects to be detected. Using this technology, a new belt of asteroid-like bodies was discovered at the edge of our Solar System in 1992. Called the Kuiper Belt, this huge reservoir of 'mini planets' represents some of the most distant objects in our Solar System.

At the centre of the Solar System is the Sun and you should appreciate that the Sun is *not* a planet, but a **star**. As such, it is a huge ball of gas, consisting mainly of the elements hydrogen and helium. At its centre, nuclear reactions release energy, which is why the Sun is hot – about 5500 °C at its surface and an amazing 5 000 000 °C at its centre. Figure 1.1 shows an impressive image of the Sun. As this book is primarily about our understanding of Mars and the possibility that life may have existed or continue to exist on that world the energy produced by the Sun is of great importance. It is also important for the Earth, without the Sun's energy the Earth would be a frozen and most probably lifeless world.

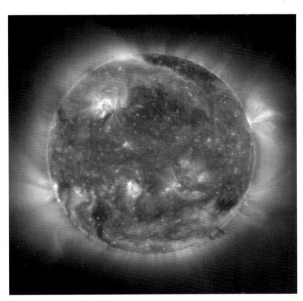

Figure 1.1 The Sun (radius 695 500 km). The image, taken in ultraviolet light using the Soho spacecraft, shows that the Sun is rather complex and 'active'.

UNDERSTANDING MARS

When considering the distances of the planets from the Sun, we inevitably find we are using huge numbers. For example, the Earth lies about 150 000 000 000 m (1.5×10^{11} m, or 150 million km) from the (centre of the) Sun. It is much more convenient to define this distance as one **astronomical unit** (AU). Thus Earth is at 1.0 AU, Jupiter at 5.2 AU and Neptune at 30.0 AU from the Sun.

> There are 1000 m (10^3 m) in a kilometre (km). See Box 3.1 in the booklet *Background material on maths skills*.

- The Earth is 149.6×10^6 km from the Sun. Mars is at 227×10^6 km from the Sun. How far is Mars from the Sun in Astronomical Units (AU)?

- To calculate the distance of Mars from the Sun in AU, you need to divide the distance in kilometres by the number of kilometres in an AU (149.6×10^6). So Mars is at $227 \times 10^6 / 149.6 \times 10^6 = 1.52$ AU.

The nine planets in the Solar System (in order of increasing distance from the Sun: Mercury, Venus, Earth, Mars, Jupiter, Saturn, Uranus, Neptune and Pluto) orbit the Sun very close to the same plane, known as the **ecliptic plane**.

> In the western world the planets are named after Roman and Greek Gods and often reflect the characteristics they show when we look at them in the night sky such as brightness, colour or the time they take to orbit the Sun.

- Are all the objects in the Solar System shown in Figure 1.2 orbiting the Sun in the ecliptic plane?

- No, the planet Pluto and comets are drawn with orbits that are not in the ecliptic plane.

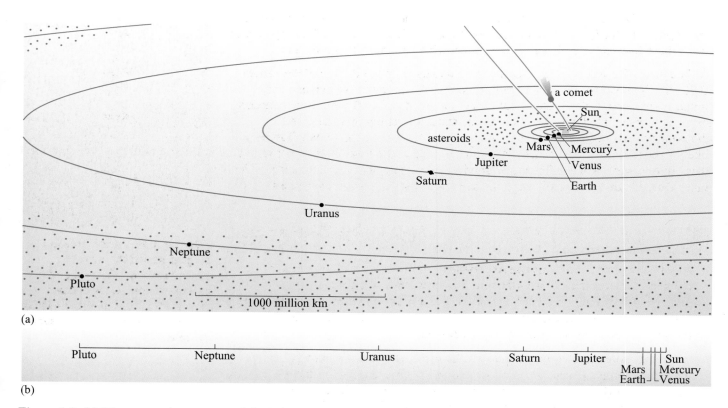

Figure 1.2 (a) Diagrammatic summary of the Solar System, showing that the planets orbit the Sun close to the ecliptic plane. Note that this is an oblique view – the orbits are actually nearly circular. (b) The correct relative distances of the planets from the Sun.

The planets orbit the Sun in elliptical orbits although eight out of the nine planets have orbits that are only very slightly elliptical, and indeed the orbits can often be approximated by a circle. Pluto is the odd one out, as its orbit is significantly elliptical. In fact Pluto's orbit takes it within the orbit of Neptune at its closest point to the Sun, and far beyond Neptune at its farthest point from the Sun (thus Pluto is the outermost planet only *some* of the time). Pluto's orbit is also significantly more inclined to the ecliptic plane than the orbits of the other planets. The sizes of the planets vary greatly, but all are dwarfed by the size of the Sun (Figure 1.3).

- ■ Look at the relative sizes of the planets in Figure 1.3. Can you group any of them together based on their size?
- ❑ There appears to be a broad division between the four small inner planets (Mercury, Venus, Earth and Mars) and the much larger outer planets (Jupiter, Saturn, Uranus and Neptune). The outermost planet, Pluto, appears to be an exception since it is a relatively small body.

We refer to the four inner planets (Mercury, Venus, Earth and Mars) as the **terrestrial planets** since they are mainly composed of rocky materials like the Earth, whereas Jupiter, Saturn, Uranus and Neptune are usually referred to as the **giant planets**.

The asteroid belt (sometimes referred to as the asteroid *main belt*) lies between Mars and Jupiter, and the Kuiper Belt lies beyond Neptune, although it should be appreciated that **minor bodies** (asteroids and comets) can be found outside these belts.

Although you have seen that the Solar System extends out to Pluto and the Kuiper Belt, strictly speaking this is not the full picture. We are aware of comets being found in our inner Solar System, but in fact the whole Solar System is surrounded by a huge spherical cloud of comets, at distances stretching up to tens of thousands of AU. This cloud, first postulated by Dutch astronomer Jan Oort, is likely to contain over 10^{11} comets that were formed in the inner Solar System, but then 'thrown out' by the influence of gravity (Box 1.1) from the giant planets. The inner **Oort cloud** occasionally supplies our Solar System with errant comets.

BOX 1.1 GRAVITY

Any two objects in the Universe experience a mutual attraction by virtue of the mass they contain. This attraction is what we call (the force of) gravity. The strength of the attraction increases in proportion to the product of the masses of the two bodies, and decreases in proportion to the square of the distance between the centres of the two bodies.

Thus, the more massive the bodies and the closer they are, the stronger the force of gravity. The Sun is by far the most massive object in the Solar System (it has a mass 357 000 times the mass of the Earth). The gravitational attraction between the Sun and the Earth holds the Earth in its orbit around the Sun, otherwise the Earth would fly off into distant space.

Figure 1.3 The relative sizes of the planets and the Sun. The planets are shown in the correct order (with increasing distance from the Sun), although the relative distance from the Sun is *not* shown to scale.

The size and period of any planet's orbit plus a knowledge of the laws of gravity enable the mass of the star (or more strictly, the combined mass of the star and the planet) to be calculated. When a planet has a relatively small satellite, measurements of the satellite's orbit about the planet allow the mass of the planet to be calculated using the same principles.

In the next two sections we'll take a quick tour around the planets of the Solar System. We'll start with a look at the remote worlds of the outer Solar System, before looking at those closer to home, the terrestrial planets, ending up on the planet we'll be examining in detail in the rest of the book: Mars.

1.2 The planets of the outer Solar System

Most of what we know about the planets of the outer Solar System comes from the detailed studies undertaken by the Voyager missions (Box 1.2) in the late 1970s and 1980s and the earlier Pioneer missions. These planets and their satellites are still the subject of detailed study by spacecraft such as the Galileo mission (Box 1.3) and the Cassini mission, which will enter orbit around Jupiter in 2004.

BOX 1.2 THE VOYAGER MISSIONS

Between them, Voyager 1 (launched 5 September 1977) and Voyager 2 (launched 20 August 1977) explored all the giant planets of our Solar System, 48 of their moons, and the unique systems of rings and magnetic fields those planets possess. Voyager 2 is still the only spacecraft to have visited Uranus and Neptune. The Voyager missions were designed to take advantage of a rare arrangement of the giant planets in the late 1970s and the 1980s that enabled fly-bys of all four planets using the minimum amount of fuel and in a relatively short time. This unique layout of Jupiter, Saturn, Uranus and Neptune occurs about every 175 years and allows a spacecraft on a particular flight path to swing from one planet to the next without the need for large onboard propulsion systems. The fly-by of each planet bends the spacecraft's flight path and increases its velocity enough to deliver it to the next destination, a procedure known as gravity assist. Using this technique, the flight time to Neptune was reduced from 30 years to 12 years.

While it was known that a mission to all four planets was possible, the expense of building a spacecraft that would accomplish such a long mission meant that the Voyager spacecraft were initially designed to conduct intensive fly-by studies of Jupiter and Saturn. The two spacecraft were sent on separate trajectories, but Voyager 2's was designed so that after encountering Saturn it would be able to fly on to Uranus and Neptune.

The Voyager interstellar mission

You may be surprised to learn that, more than 25 years after they were launched, the Voyager mission is still ongoing. After its fly-by of Saturn in November 1980, Voyager 1 was on a trajectory that took it above the ecliptic plane of the Solar System. Similarly, after Voyager 2's fly-by of Neptune the spacecraft headed below the ecliptic plane. In 1998, the distance from the Sun of Voyager 1 exceeded that of the earlier Pioneer 10 spacecraft making Voyager 1 the most distant human-made object in space. It is currently leaving the Solar System at a speed of about 3.6 AU per year (about 38 000 miles per hour) and as of December 2003 is 90 AU from the Sun. Both spacecraft continue to return valuable scientific data on the extent of the Sun's magnetic field and other properties as they approach the space between the stars: interstellar space.

1.2.1 Jupiter

Jupiter is the largest planet in the Solar System (see Figure 1.3) with a radius of over ten times that of Earth and a mass of about three hundred times that of Earth. Its density is about one-quarter of that of Earth; the reason for this is that Jupiter consists mostly of gas and (along with Saturn which has similar properties) is sometimes referred to as a **gas giant**.

The most striking features of Jupiter are the colourful bands and the swirling clouds. The study of these clouds was one objective of the first spacecraft to cross the asteroid belt, Pioneer 10, which was launched in 1972 and flew by Jupiter in December 1973. What you are seeing in Figure 1.4 is the top of a continuously moving gaseous atmosphere that extends deep into the planet. The whole planet rotates in just 9.9 hours, although the atmosphere at the poles and the equator rotates at slightly different rates, giving rise to 'winds' of 200 m s^{-1} (around 450 mph) at the equator. The largest feature is the Great Red Spot, shown in Figure 1.4, with a close-up shown in Figure 1.5 taken by one of the Voyager spacecraft (Box 1.2), which is around 20 000 km across. It is a huge storm that has been observed for the last few hundred years, continuously rotating in an anti-clockwise direction with wind speeds of around 100 m s^{-1} at the edges. The atmosphere itself is 90% hydrogen and 10% helium, with traces of methane, ammonia and water vapour. The cloud-top temperature is around 150 °C, although beneath the surface the temperature and pressure must rise rapidly due to the compression of the atmosphere by Jupiter's immense gravity. At a depth of 10 000 km, the pressure is likely to be a million times that on Earth, with a temperature of 6000 °C (approximately the temperature of the surface of the Sun). At Jupiter's core, the pressure is likely to be an amazing 40 million times that on Earth, and the temperature 16 000 °C.

The many satellites of Jupiter (there are at least 61) offer an incredibly diverse selection of planetary bodies and have been the subject of detailed study by the Galileo mission (Box 1.3). Figure 1.6 shows an image of Io, a satellite of Jupiter. Its incredible surface (sometimes described as pizza-like in appearance) shows evidence of a vast amount of volcanic activity. The source of energy for this volcanic activity is Jupiter itself. Io is the innermost major satellite and, in much the same way as the Moon causes tides on Earth, Jupiter distorts Io as it orbits the planet causing Io to heat up in a process known as tidal heating. Io is mainly 'rocky' in composition and temperatures in the region of 700 °C and more are required to melt these rocky materials.

Figure 1.4 The planet Jupiter (radius 69 910 km). The Cassini spacecraft took the image in 2000, while en route to Saturn. The Great Red Spot is by far the largest feature of the planet.

Figure 1.5 Jupiter's Great Red Spot. The spot is about 20 000 km across.

Figure 1.6 Io (radius 1821 km), a satellite of Jupiter. The Galileo spacecraft took this image.

Impact craters are formed when asteroids and comets slam into the surface of a planet at great speed.

The possibility of life on Europa, however remote, is the reason NASA decided to prevent the risk of the Galileo spacecraft colliding with Europa by destroying the spacecraft in its impact with Jupiter on 21 September 2003.

The next satellite we consider could hardly be more of a contrast to Io. Europa (Figure 1.7) has an icy surface that is covered in cracks, with hardly any **impact craters**. As with Io, the lack of craters is an indication of a relatively young surface. Europa is a predominantly rocky body. However, it has a layer of ice, about 100 km deep, on the surface. Europa would also be expected to undergo significant heating due to the influence of Jupiter's gravity (albeit to a much lesser extent than Io as it is farther from Jupiter). This presents a fascinating possibility that beneath Europa's icy 'crust' there is a global ocean of melted ice (i.e. water). This also leads to speculation that, where there is liquid water, there might be primitive life!

The possibility of an icy crust on top of a subsurface ocean does appear consistent with the surface features on Europa, which look remarkably like broken ice packs and fractured ice plains (Figure 1.7). The surface of Europa also leads us to a new concept – that of **cryovolcanism**. This is the name given to the effect where cold slurries of ice and liquid erupt and flow across the surface like 'cool volcanoes', just as hot molten rock erupts and flows as lava. It appears that cryovolcanism is commonplace amongst the icy satellites of the giant planets, and the surface of Europa is one such case. The relatively recent activity has wiped clean the impact craters that would otherwise have been visible on Europa's surface.

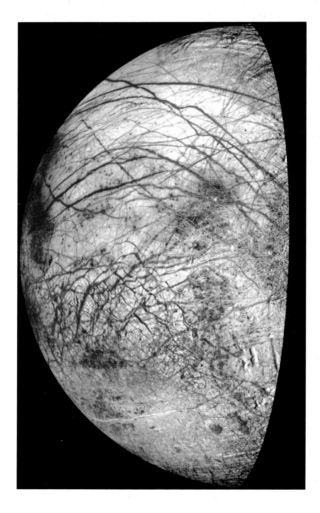

Figure 1.7 Europa (radius 1565 km), which has been in orbit around Jupiter since 1995, appears as a crescent in this enhanced-colour image taken by the Galileo spacecraft.

Jupiter has two even larger satellites than Io and Europa. These are Ganymede (radius 2634 km) and Callisto (radius 2403 km). Ganymede is the largest satellite in the Solar System and is actually bigger than the planet Mercury. However, it contains much less mass than Mercury because, like Callisto, it is a predominantly icy body. Both these large satellites are heavily cratered, showing that their surfaces are much more ancient than those of Io and Europa.

BOX 1.3 THE GALILEO MISSION TO JUPITER

The Galileo spacecraft, named after the Italian astronomer Galileo Galilei who discovered the four largest moons of Jupiter (Io, Europa, Ganymede and Callisto), was launched on 18 October 1989, becoming the first spacecraft to go into orbit around a giant planet in December 1995. Galileo also carried a probe that became the fist spacecraft to directly sample the atmosphere of a giant planet. Galileo's primary mission was to study Jupiter's atmosphere, its magnetic field, and the four largest moons for two years (1995–1997). In an extended mission from 1997–1999, Galileo studied in further detail and closer range Jupiter's icy moon Europa and its volcanic moon Io. The mission finally came to an end on 21 September 2003 when the spacecraft impacted into Jupiter's atmosphere to prevent any possible future contamination of Jupiter's moons.

Galileo's probe entered Jupiter's atmosphere and measured temperature, pressure, chemical composition, cloud characteristics, sunlight and energy internal to the planet, and lightning, all during its brief 59-minute descent. It penetrated 200 km into the violent atmosphere of Jupiter – measuring winds of 450 mph – before it was crushed, melted, and/or vaporized by the pressure and temperature of the atmosphere.

1.2.2 Saturn

The next planet out is Saturn (Figure 1.8), instantly recognizable because of the prominent system of rings around it. Saturn itself is not quite as massive as Jupiter, being 'only' 95 Earth masses and about 15% smaller than Jupiter in radius. The clouds seen in Figure 1.8, with somewhat enhanced colours, form bands across the disc that are parallel to its equator. They are rather like the bands and clouds of Jupiter, although obvious storms and swirling structures are not apparent. The planet rotates in just 10.7 hours. This gives rise to the atmosphere bulging at the equator. You may have noticed in Figure 1.8 that the planetary disc does not appear to be perfectly circular, but is somewhat flattened. This is a real effect and not just a distortion in the image. Winds on the equator reach speeds of around 500 metres per second, and storms and even 'spots' can evolve (although not to the extent seen on Jupiter).

The rings of Saturn are remarkable, and beautiful. They are not solid, but are made of icy particles and boulders, the majority of which are between about a centimetre and a few metres in size. These particles most probably originate from the catastrophic break-up of a satellite due to impact. The ring particles all orbit the planet in the same plane (Saturn's equatorial plane), creating an amazingly thin disc – the rings are only about tens of metres thick yet are over 200 000 km across.

Figure 1.8 The planet Saturn (radius 58 230 km). This Voyager image has had the colours enhanced slightly to make the banding in the clouds more obvious. The apparent break in the rings just to the right of the planetary disc is simply where the rings are in shadow.

The ring particles, being icy, are highly reflective. This makes them easy to see in reflected sunlight. If you could gather up all the particles in the entire ring system, and put them together into one object, this object would be about 3000 km in diameter. Figure 1.9 shows a close-up image of the rings. Notice that there are distinct 'gaps' in the rings. These arise because of the gravitational influence of Saturn's satellites 'shepherding' the ring particles.

Saturn has over thirty satellites, but only eight can be classed as major (i.e. those with radii greater than about 200 km). One of these, Titan (Figure 1.10), is one of the larger satellites in the Solar System, being a little under half the size of Earth. What makes it very interesting is that it has a thick, predominantly nitrogen, atmosphere – a description that also fits Earth. Although the exact composition is not yet known, it is likely that nitrogen accounts for 82–99% by volume (Earth's atmosphere is 78% nitrogen by volume), with the rest being mainly methane and other hydrocarbons (compounds made of hydrogen and carbon), which give Titan its orange colour. Because the atmosphere is opaque, we have very little information about the surface. It is likely to be icy, with a surface temperature of around −180 °C), and an atmospheric pressure about 1.5 times that of Earth. It is possible that methane and other hydrocarbons in the atmosphere could condense into droplets, producing a hydrocarbon rain. This would give rise to lakes or seas of liquid methane and liquid hydrocarbons. One can imagine that the surface might bear some resemblance to a deep-frozen Earth.

Figure 1.9 The rings of Saturn. This Voyager image shows clearly the gaps between the rings.

Figure 1.10 Titan (radius 2575 km), a satellite of Saturn. Titan is potentially one of the most interesting satellites in the Solar System although, as seen in this Voyager image, its thick atmosphere makes it look featureless and somewhat unexciting.

The European Space Agency (ESA) Huygens probe, a small spacecraft about one metre across carried aboard the NASA Cassini spacecraft, is designed to answer many questions about Titan. Upon arrival at Saturn in 2004, the Huygens probe will be released from the Cassini spacecraft in order to enter the atmosphere of Titan and make a 2-hour descent to the surface by parachute. Various instruments on board Huygens are designed to measure the properties of the atmosphere and the icy, rocky or liquid nature of the surface, thus allowing us to learn a great deal about Titan.

NASA stands for National Aeronautics and Space Administration.

Other large satellites of Saturn include Enceladus (radius 249 km), Tethys (radius 530 km), Dione (radius 560 km), Rhea (radius 764 km) and Iapetus (radius 718 km).

1.2.3 Uranus

Uranus is smaller than both Jupiter and Saturn. Uranus has a mass of almost fifteen times that of Earth and is about four times larger in size than Earth. Sir William Herschel discovered the planet in 1781. Images show a rather featureless blue–green planet, as shown in Figure 1.11. One oddity of Uranus is that its spin axis is tilted over 98 degrees from the 'vertical'. This is most likely due to a huge impact event early in the planet's history that literally knocked Uranus over on its side. It means that one pole of the planet can point towards the Sun for long periods of time. The resulting polar heating drives the atmospheric flow on the planet, with winds flowing from the poles to the equatorial regions. The prominent banding and storms seen on Jupiter and Saturn are not seen on Uranus, although some subtle structure is exposed in some false-colour images.

Figure 1.11 The planet Uranus (radius 25 560 km). This image is from Voyager 2, taken in 1986. The planet has a rather featureless blue–green atmosphere.

Uranus, like Saturn, has a multitude of satellites (at least twenty and probably more), although only five are considered major bodies: Miranda (radius 236 km), Ariel (radius 579 km), Umbriel (radius 585 km), Titania (radius 789 km) and Oberon (radius 761 km).

1.2.4 Neptune

Neptune (Figure 1.12) has a mass of about seventeen times that of the Earth and is only slightly smaller than Uranus (being about four times larger than Earth). The existence of Neptune was suspected for some time, after detailed observations of the motion of Uranus suggested that there must be another large planet in the outer Solar System gravitationally influencing the orbit of Uranus. Astronomers Johann Galle and Heinrich D'Arrest discovered Neptune in 1846. Its atmosphere has an intense blue colour. More features are apparent in the atmosphere than for Uranus, with some banding and pale clouds being evident. Neptune also appears to have its own version of Jupiter's Great Red Spot. Neptune's 'Great Dark Spot' is also an oval-shaped storm system, although it is much less long-lived than Jupiter's spot (Hubble Space Telescope images reveal that the Great Dark Spot has already dispersed).

Figure 1.12 The planet Neptune (radius 24 770 km). This Voyager 2 image, showing a 'Great Dark Spot', was obtained in 1986.

The Hubble Space Telescope is an optical telescope placed in Earth orbit in 1990.

Neptune has at least eight satellites, although only three are of any great size, Proteus (radius 209 km), Triton (radius 1353 km), and Nereid (radius 170 km). Triton (Figure 1.13) is Neptune's only large satellite. The surface is an icy mixture of nitrogen, carbon monoxide, methane and carbon dioxide. It has a temperature of −230 °C). One of the unusual things about Triton is that it orbits Neptune in the opposite direction to most other satellites. This is evidence that Triton was captured rather than forming near Neptune from 'leftover' material. This capture process will probably have involved violent impacts with other existing satellites, and will have left Triton in an orbit that gave rise to significant tidal heating. This all means that the surface of Triton has undoubtedly had a complex geological history, and that the surface is probably heavily modified from its original appearance.

Figure 1.13 Triton (radius 1353 km), a satellite of Neptune. Triton is Neptune's only large satellite.

The terrain on Triton appears divided. The area on the left of Figure 1.13 is Triton's southern polar cap. Some dark streaks are apparent in this region. These are due to cryogenic geyser-like eruptions that send plumes of dust about 10 km above the surface, and then leave dark stains across the surface. These geyser plumes were actually seen by Voyager 2, showing that activity is ongoing. This sort of activity gives rise to a very tenuous nitrogen atmosphere. The right-hand part of the Figure 1.13 shows a different texture, which has been likened to the skin of a cantaloupe melon. Cryogenic lavas have probably flooded much of this region. Triton has probably got much in common with Pluto (the last planet on our tour of the outer planets) and will have originated from the same region of the outer Solar System.

QUESTION 1.1

Which *one* of the following statements about the giant planets and their satellites is *incorrect*?

A The lack of impact craters on Europa indicates that this satellite has a relatively young surface.

B The axis of rotation of the planet Uranus is tilted over so that the poles of the planet point toward the Sun.

C The atmosphere of Jupiter is composed of the gases hydrogen and nitrogen.

D Some of the gases in Titan's atmosphere (a satellite of Saturn) contain compounds made of hydrogen and carbon.

E Cryovolcanism has operated on Triton.

1.2.5 Pluto

Pluto, the last planet on our tour, is the only one that has not yet had a spacecraft mission fly near it. For this reason we know less about it than the other planets. However, we do know that it would be an extremely interesting place to visit. Astronomer Clyde Tombaugh discovered Pluto in 1930 after a huge photographic search. This little planet offered a big surprise in 1978 when a satellite, named Charon, was discovered. However Charon is about half the size of Pluto, and so the system is not so much a planet with a satellite, as a binary planetary system (or 'double planet'). Figure 1.14 shows a Hubble Space Telescope image of the Pluto–Charon system. The bodies orbit each other, separated by about 19 400 km. Although this image does not show any detail on the bodies, some important properties have been determined by various other means. Both rotate such that the same hemispheres face each other all the time. The Charon-facing side of Pluto is a lot brighter (more reflective) than the other side. Pluto also has a very tenuous (mainly nitrogen) atmosphere, similar to that seen on Triton. Indeed, it is likely that Pluto is quite similar to Triton (although less geologically processed). Both Pluto and Charon are believed to have formed beyond Neptune in the region called the **Kuiper Belt**. The Kuiper Belt refers to a belt of planetary bodies beyond Neptune, of which Pluto appears to be the largest member.

Figure 1.14 Pluto (radius 1137 km) and its satellite Charon (radius 586 km). This image was obtained using the Hubble Space Telescope and shows the two bodies 19 400 km apart.

Kuiper is pronounced KY-per with the emphasis on the first syllable.

If Pluto was formed in the Kuiper belt, is it really a planet? The answer is probably no. However, since Pluto was regarded as a planet when it was discovered in 1930 and this status remained unchallenged for 50 years until the discovery of Kuiper belt objects, most astronomers are content to keep its official status as a planet, for historical reasons.

1.3 The planets of the inner Solar System

We now turn our attention to the planets of the inner Solar System: the terrestrial planets Mercury, Venus, Earth, and of course Mars. Aside from their smaller size, these planets differ from those of the outer Solar System in their composition, they are primarily rocky as opposed to the large amounts of gas that make up Jupiter, Saturn, Uranus and Neptune.

1.3.1 Mercury

Mercury (Figure 1.15), being the closest planet to the Sun, can get very hot. In sunlight, parts of the surface can reach about 470 °C, whereas in darkness the temperature can drop to about −190 °C). Looking at Figure 1.15, perhaps the most striking features are the round 'scars' on the surface. These are impact craters, which are formed when asteroids and comets slam into the surface of a planet at great speed. Any undisturbed surface of a planet will accumulate impact craters over time. Thus a very cratered surface implies that the surface is relatively old, whereas a lack of craters might indicate that the surface has been renewed or replaced in some way, wiping out the craters from the surface.

Figure 1.15 The planet Mercury (radius 2440 km). The Mariner 10 spacecraft obtained this image in 1974 (so far the only spacecraft to go to Mercury). The surface is covered in round 'scars' known as impact craters formed when objects such as asteroids or comets have slammed into Mercury's surface.

■ Can you think of a naturally occurring process on Earth that replaces or resurfaces parts of the planet?

❑ The most prevalent mechanism for resurfacing is the action of volcanoes, whereby lava (the melted rock we are familiar with on Earth) flows and covers a pre-existing surface.

On Mercury, the cratering over the surface appears reasonably uniform, although there are some relatively small areas that look smoother, indicating some volcanic resurfacing has taken place. You can see from the clarity of the image in Figure 1.15 that Mercury does not have an obscuring atmosphere. In fact, Mercury does have some extremely tenuous atmosphere, but it is a thousand million million (i.e. 10^{15}) times less dense than the atmosphere on Earth!

1.3.2 Venus

The next planet out from the Sun on our tour is Venus. The chances are that you have seen Venus with the naked eye, even if you didn't realize it at the time. Venus is often seen as an extremely bright 'star' an hour or two before sunrise or after sunset, depending on the relative positions of Venus and the Earth in their orbits. When viewed through a telescope, Venus appears as a featureless planet due to the presence of a thick atmosphere. In terms of its size, and the fact that it has a significant atmosphere, Venus could be considered as the 'twin' of Earth. In fact there are very important differences, particularly regarding the composition of the atmosphere and the resulting surface environment. Figure 1.16 shows an image of Venus, which picks out some cloud structure that is not normally apparent. The clouds are made from tiny droplets of sulfuric acid, hinting that Venus might not be the most welcoming environment for us to visit!

Figure 1.16 The planet Venus (radius 6052 km). This image, taken by the Galileo spacecraft, is falsely coloured to highlight the subtle structure of the clouds that are not usually seen (normally Venus looks more of a uniform white in appearance).

A view of the surface terrain can be obtained using an instrument based on the principles of radar whereby radio waves are aimed through the clouds whilst their reflections back from the surface are monitored. One such instrument was carried by the Magellan spacecraft (Box 1.3). One such image is shown in Figure 1.17. The surface of Venus is very complex, with far fewer impact craters than on Mercury, but with many volcanoes and **lava** plains suggesting significant surface renewal. The only images obtained from the surface of Venus were taken from a series of Soviet Union spacecraft, called Venera. Taking images of Venus was an impressive technical feat considering the hostility of the surface environment. The surface atmospheric pressure was almost a hundred times that on Earth, and the temperature was around 400 °C. A high-pressure oven is not a good place for sensitive scientific instruments! However before the equipment expired, the Venera spacecraft returned their precious images. Figure 1.18 shows one of the few colour images obtained by the Venera 13 spacecraft. The atmosphere of Venus is mostly (97% by volume) carbon dioxide (unlike Earth which is mostly nitrogen and oxygen). The carbon dioxide gives rise to a strong 'greenhouse effect' that traps heat below the lower layers of the atmosphere – hence the very high surface temperature. Venus, while an Earth-twin in some respects, would definitely not be a hospitable place to visit.

The 'greenhouse effect' raises the temperature at a planet's surface as a result of heat energy being trapped by gases such as carbon dioxide in the atmosphere.

Figure 1.17 Details of the surface of the planet Venus, which is usually totally obscured by clouds, taken by the Magellan spacecraft using cloud-penetrating radar.

Figure 1.18 The surface of Venus imaged by the Venera 13 spacecraft in 1982. (Part of the spacecraft is seen at the bottom of the image.)

BOX 1.3 MISSIONS TO VENUS

In December 1962, Venus was the first planet in the Solar System to be visited by a spacecraft from Earth when Mariner 2, which had been launched in August that year, flew within 35 000 kilometres of the Venusian surface and discovered a planet obscured by perpetual cloud cover. Venus was subsequently visited by other NASA spacecraft including Pioneer Venus in 1978 that carried probes to examine the planet's atmosphere and the series of Soviet Venera spacecraft and landers, which probed the planet in the 1970s and early 1980s.

However, most of our detailed knowledge of the surface and interior of Venus comes from the NASA Magellan mission, which was launched from a space shuttle in May 1989, entering orbit around Venus in August 1990. Magellan's primary task was to map the surface of Venus using radar in order to study its landforms, geological processes such as erosion, deposition, impact, and chemistry at work on the surface; and to model the interior of the planet by studying variations in its gravity field. In all, Magellan mapped 98% of the planet's surface and found a planet dominated by volcanoes, which have left their marks on 85% of Venus in the form of large volcanoes, lava plains, and extremely long lava channels (Figure 1.19).

Figure 1.19 A 3-dimensional reconstruction of the Venusian volcano Maat Mons obtained from the radar maps acquired by the Magellan spacecraft. The image shows the extensive lava flows that extend for hundreds of kilometres from the volcano across the fractured plains shown in the foreground.

1.3.3 Earth

The familiar blue planet is shown in Figure 1.20. Although the Earth may seem rather familiar, it is worth pausing for thought when looking at Figure 1.20. Consider just how important this image is for showing us our home planet and its place in the Solar System. Before the space age, we could only imagine seeing our planet from afar. But now we have an appreciation of the Earth as a finite, isolated and even rather fragile planet in space. More remarkable still is the image shown in Figure 1.21, which was taken in May 2003 by the Mars Global Surveyor spacecraft in orbit around Mars. It shows how the Earth and the Moon appear to an observer on Mars with a relatively low powered telescope!

The Earth also allows us to study at close quarters many of the processes that influence and characterize the other planets in the Solar System. Our understanding of the internal structures of planets such as Mars and Venus, processes such as volcanism and the composition of atmospheres, is greatly enhanced by looking at what happens on (or *in*) the Earth, and using this knowledge to consider what must happen elsewhere. For this reason much of the material considered in the following chapters uses knowledge gained from the detailed study of the Earth to enhance our understanding of Mars.

> The atmosphere of the Earth is crucial for the survival of life on the planet. It is responsible for a significant rise in surface temperature because of a 'greenhouse effect'(modest compared to Venus, but still accounting for a 33 °C higher temperature than an atmosphereless Earth would have), which is mainly due to carbon dioxide and water vapour. This means that the *mean* temperature at the surface of Earth is 15 °C, allowing liquid water to exist over much of the planet.

Figure 1.20 The planet Earth (radius 6378 km). This image was taken from Apollo 16 in 1972. The image shows the oceans (blue), land (brown) and cloud (white) as well as the Antarctic ice-cap (uniform white at the bottom of the figure).

Figure 1.21 The Earth and moon seen from Mars. In May 2003, the Mars Global Surveyor spacecraft in orbit around Mars took this remarkable image of the Earth and Moon. At the time, Mars was 139 million km from Earth.

The atmosphere also carries heat away from the Equator, so that the Equator is not as hot as it might be and the polar regions are not as cold as they might be. This allows life to survive at a greater range of latitudes than would otherwise be the case if we didn't have atmospheric circulation. The Earth's atmosphere comprises (by volume) 78% nitrogen and 21% oxygen, with other gases (including carbon dioxide) being just a small part. It is perhaps a sobering thought to bear in mind that the 400 °C surface temperature of Venus is an example of what could happen if greenhouse gases such as carbon dioxide became a really significant proportion of the Earth's atmosphere.

The other familiar 'planetary body' on our tour is the Earth's only natural satellite, the Moon. Often ignored due to over familiarity, the Moon offers a spectacular (if somewhat grey-coloured) terrain. Viewed with the naked eye or, better still, through binoculars or a small telescope, the relatively bright 'highland' regions peppered with impact craters, and the darker and less cratered *mare* (pronounced mar-ray) regions, are clearly seen.

The Moon orbits the Earth about thirteen times each year, and presents the same face to us all the time (Figure 1.22). This is because the time it takes to turn once on its axis exactly matches the time it takes to travel once around the Earth. Figure 1.23 shows a photograph of the far side of the Moon. Fewer *mare* regions are seen and more impact craters are obvious in this image. The mare regions are younger formations formed by floods of lava that wiped away ancient impact craters.

Figure 1.22 The Moon (radius 1738 km). The Clementine spacecraft took this image of the nearside of the Moon.

Figure 1.23 An image of the far side of the Moon taken by the Clementine spacecraft.

1.3.4 Mars

And so we come to Mars, the subject of this book and the planet in the Solar System that will be intensively studied by spacecraft over the next few years. The advent of the era of exploration of Mars by interplanetary spacecraft brought an abrupt end to many of the more speculative ideas about the planet that you met in the *Introduction*.

Figure 1.24 The planet Mars (radius 3394 km). The view of Mars shown was assembled from individual images obtained by the Mars Global Surveyor spacecraft on 12 May 2003. At that time, the northern hemisphere was in early autumn and the southern hemisphere in early spring.

Each successive visit to the planet by **fly-by**, **orbiter** and **lander** missions has resulted in a radically new picture of Mars emerging and old misconceptions have been laid to rest. The first successful missions to Mars were by the Mariner spacecraft, Mariners 4, 6 and 7 (Section 2.2), which flew by the planet at close range in the 1960s and returned the first close-up images of its surface. They suggested that Mars was a barren place covered with craters and apparently lifeless, similar to the Moon. However, our knowledge of Mars was changed again with the arrival at Mars of Mariner 9 in November of 1971. This spacecraft was an orbiter, not a fly-by mission like its predecessors, and it began a systematic exploration that showed Mars to be a complex planet, very unlike the Moon.

The Mars Global Surveyor spacecraft in orbit around the planet took the image of Mars shown in Figure 1.24. It is not hard to understand why Mars is often referred to as 'the red planet' (although most people would probably describe it as orange). Mars can often be seen with the naked eye as a 'star' that has a very obvious orange hue to it. Figure 1.24 shows some of the striking features that have been revealed by the exploration of Mars by spacecraft. The linear feature in the centre of the image shows a huge canyon system (called *Valles Marineris*), which represents a fracture in the planet's surface that extends about 4000 km across the planet. The conventions used for naming surface features on Mars and other planets are outlined in Box 1.4. This canyon dwarfs the Earth's Grand Canyon, having regions that are 11 km deep and 200 km wide. Also very obvious are the circular features near the left-hand side of the image. These are enormous, old volcanoes. The largest volcano on Mars, *Olympus Mons* (Figure 1.25), which is also the largest volcano in the Solar System, is 24 km high and has a volume one hundred times greater than Mauna Kea in Hawaii – the largest equivalent feature on Earth.

Figure 1.26 shows an image of the surface of Mars taken by the Mars Pathfinder lander in 1997. Large boulders embedded in dust and 'soil' can be seen in the rather barren landscape. Lava flows have altered the terrain in other regions of the planet, and even evidence of ancient running water has now been identified. The atmosphere of Mars is mainly carbon dioxide (95% by volume) and in this respect it is similar to the atmosphere of Venus. However, on Mars the atmospheric pressure at the surface is much reduced, being only about 0.006 times that on Earth.

> Mars' rather tenuous atmosphere means that, although there is a high percentage of carbon dioxide, the greenhouse effect it produces is very modest, adding only about 6 °C to the mean temperature of the planet (which is −50 °C).

The surface of Mars is desert-like – it is very dry and can get relatively warm during the day and extremely cold at night (see Section 5.2). Although the atmosphere is thin, winds on the planet can be formidable, giving rise to large dust storms that can last for weeks or months at a time.

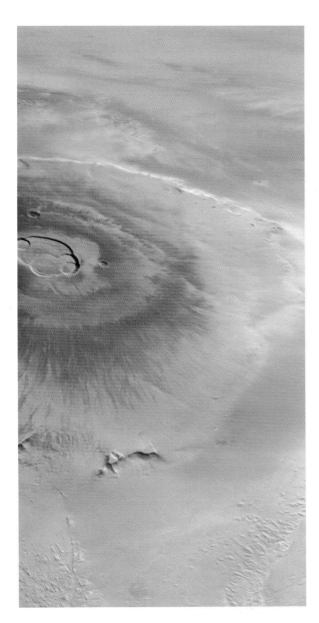

Figure 1.25 The giant volcano Olympus Mons on Mars. The volcano is over 24 km high and 600 km in diameter – three times higher than Mount Everest – and as wide as the entire Hawaiian Island chain. The Mars Global Surveyor spacecraft took this image.

Figure 1.26 The surface of the planet Mars, as imaged by the Mars Pathfinder lander mission in 1997.

BOX 1.4 WHAT'S IN A NAME?

The International Astronomical Union (IAU) is the international body that oversees many aspects of astronomy and planetary science, in particular as regards discoveries of new astronomical objects. The IAU has a set of naming conventions for naming the surface features of planets and other planetary bodies. These specify themes for the names, for example on Mars large craters are named after deceased scientists who have contributed to the study of Mars or writers and others who have contributed to the lore of Mars. Large channels are named after the word for Mars in various languages.

Names for all planetary features include a descriptor term, with the exception of craters where the descriptor term is implicit. Descriptor terms are simply descriptions of what a feature looks like and are not intended to give any information about how the feature was formed. Some of the more common descriptor terms you may meet in your exploration of Mars are listed in Table 1.1.

Table 1.1 Some of the common descriptor terms and meanings for features on planetary surfaces.

Descriptor term (plural in brackets)	Meaning
Albedo (Albedos)	geographic area distinguished by amount of reflected light
Catena (Catenae)	chain of craters
Cavus (Cavi)	hollows, irregular steep-sided depressions usually in clusters
Chaos	distinctive area of broken terrain
Chasma (Chasmata)	a deep, elongated, steep-sided depression
Colles	small hills or knobs
Corona (Coronae)	ovoid-shaped feature
Crater (Craters)	circular depression
Dorsum (Dorsa)	ridge
Fluctus (Flucti)	flow terrain
Fossa (Fossae)	long, narrow, shallow depression
Labes (Labes)	landslide
Labyrinthus (Labyrinthi)	complex of intersecting valleys
Linea (Lineae)	a dark or bright elongate marking, may be curved or straight
Macula (Maculae)	bright spot
Mesa (Mesae)	flat-topped prominence with cliff-like edges
Mons (Montes)	mountain
Patera (Paterae)	irregular crater, often with scalloped edges
Planitia (Planitiae)	low-lying plain
Planum (Planae)	plateau or high plain
Regio (Regiones)	a large area marked by reflectivity or colour distinctions from adjacent areas, or a broad geographic region
Rupes (Rupes)	scarp
Scopulus (Scopuli)	lobe-shaped or irregular scarp
Sulcus (Sulci)	parallel or nearly parallel ridges and furrows
Tholus (Tholi)	small dome-like mountain or hill
Unda (Undae)	dunes (usually used only in plural)
Vallis (Valles)	valley
Vastitas (Vastitates)	extensive plain

Phobos, Deimos and the asteroid belt

Mars has two relatively tiny satellites, Phobos and Deimos, which are thought to be asteroids that have been captured by the gravitational influence of Mars. The bodies are irregularly shaped: Phobos is approximately 26 km × 18 km in size, Deimos is approximately 16 km × 10 km in size. Phobos is shown in Figure 1.27, and you can see a large (relative to the size of the body) impact crater on the left-hand side, as well as other smaller impact craters. Phobos orbits only 6000 km above the surface of Mars, and will probably collide with Mars within the next 50 million years. Deimos is shown in Figure 1.28. There are few craters seen, and the surface may be covered in fine dust.

Phobos and Deimos are tiny compared to the Moon and to the satellites of the planets of the outer Solar System.

Between Mars and Jupiter is the **asteroid belt**, which is a 'swarm' of rocky and metallic bodies. The asteroid belt extends all the way around the Sun and each asteroid orbits the Sun. It is from this reservoir of bodies that Phobos and Deimos may have originated.

Figure 1.27 Phobos (26 km × 18 km), a satellite of Mars. This body is thought to be a captured asteroid.

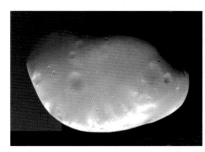

Figure 1.28 Deimos (16 km × 10 km), a satellite of Mars. As with Phobos, this body is also thought to be a captured asteroid.

Asteroids do not always stay in the asteroid belt. Due to collisions with each other and the gravitational effects of Jupiter or Mars their orbits can change. Some enter orbits that bring them into the inner parts of the inner Solar System crossing the orbits of Mars, Earth, Venus and Mercury and, on occasions, colliding with these planets to produce impact craters.

1.4 Summary of Chapter 1

- The Solar System is broadly separated into terrestrial planets (predominantly rocky and metallic in composition) in the inner region, and giant planets (predominantly gaseous) in the outer region.
- The planets orbit the Sun on elliptical orbits, although the orbit of each planet, with the exception of Pluto, can be approximated to a circle.
- The planets, in order of increasing distance from the Sun, are Mercury, Venus, Earth, Mars, Jupiter, Saturn, Uranus, Neptune and Pluto (although Pluto is closer to the Sun than Neptune for part of its orbit).
- The distance between the Sun and the Earth (more correctly the mean distance between their centres) is defined as one astronomical unit (1 AU).
- There is a belt of asteroids (rocky and/or metallic minor bodies) lying mainly between the orbits of Mars and Jupiter (although asteroids can be found elsewhere in the Solar System albeit in smaller numbers).
- There is another belt of minor bodies lying beyond Neptune called the Kuiper Belt.
- Comets (icy minor bodies) can be found throughout the Solar System (particularly the inner Solar System). Comets also make up the Oort cloud (a massive spherical shell of comets at great distances from the inner Solar System).

- Images of planetary bodies reveal a great diversity in their surfaces. Impact craters (made by asteroids and comets) are common on most bodies. A lack of impact craters implies that there has been some resurfacing (the surface is relatively young).
- Evidence of volcanism and lava flows can be seen on the terrestrial planets, and evidence of cryovolcanism can be seen on many of the icy satellites.

1.5 Questions

QUESTION 1.2

Using Table 1.1 as a guide, what sort of terrain are the following martian features describing. (a) Isidis Planitia, (b) Syrtis Major Planum, (c) Gordii Dorsum, (e) Orcus Patera?

QUESTION 1.3

Summarize the main features of (a) the atmosphere, (b) surface temperature, (c) surface topography of the planets Mercury, Venus, Earth and Mars.

QUESTION 1.4

Which *two* of the following statements about Mars are *incorrect*?

A Lowell would be a suitable name for a martian impact crater.

B Mars has an atmosphere that contains a higher percentage (by volume) of carbon dioxide than Earth's atmosphere.

C The martian volcano Olympus Mons has a volume 100 times greater than the largest terrestrial volcano.

D Valles Marineris is a canyon on Mars, similar in size to the Grand Canyon on Earth.

E Mariner 2 was the first spacecraft to visit Mars.

QUESTION 1.5

Briefly describe the main contribution of Giovanni Schiaparelli to the study of Mars.

CHAPTER 2
MISSIONS TO MARS

2.1 Introduction

Up until the end of the year 2003, over 30 space missions have attempted to explore Mars (Table 2.1). Starting with an unsuccessful Soviet mission (official designation Mars 1960A), which failed on launch in 1960, the space exploration of Mars has been dogged by many failures, especially in the early days of the space age.

Table 2.1 Spacecraft Missions to Mars ordered by date of launch.

Mission	Launch date	Remarks
Mars 1960A (USSR)	10 Oct 1960	attempted fly-by; launch failure
Mars 1960B (USSR)	14 Oct 1960	attempted fly-by; launch failure
Mars 1962A (USSR)	24 Oct 1962	attempted fly-by; failed to leave Earth orbit
Mars 1 (USSR)	1 Nov 1962	fly-by; lost contact in transit
Mars 1962B (USSR)	4 Nov 1962	attempted lander; failed to leave Earth orbit
Mariner 3 (USA)	5 Nov 1964	attempted fly-by
Mariner 4 (USA)	28 Nov 1964	fly-by, imaging
Zond 2 (USSR)	30 Nov 1964	fly-by, lost contact in transit
Mariner 6 (USA)	24 Feb 1969	fly-by, imaging and atmospheric measurements
Mariner 7 (USA)	27 Mar 1969	fly-by, imaging and atmospheric measurements
Mars 1969A (USSR)	27 Mar 1969	attempted lander; launch failure
Mars 1969B (USSR)	2 Apr 1969	attempted lander; launch failure
Mariner 8 (USA)	8 May 1971	attempted lander; launch failure
Cosmos 419 (USSR)	10 May 1971	attempted orbiter/lander
Mars 2 (USSR)	19 May 1971	orbiter; lander crashed on surface
Mars 3 (USSR)	28 May 1971	orbiter; lander lost contact
Mariner 9 (USA)	30 May 1971	orbiter; imaging of Mars, Phobos and Deimos
Mars 4 (USSR)	21 July 1973	fly-by imaging; attempted orbiter
Mars 5 (USSR)	25 July 1973	orbiter; imaging
Mars 6 (USSR)	5 Aug 1973	orbiter; lander lost contact, some data
Mars 7 (USSR)	9 Aug 1973	orbiter; attempted lander
Viking 1 (USA)	20 Aug 1975	orbiter and lander in Chryse Planitia
Viking 2 (USA)	9 Sept 1975	orbiter and lander in Utopia Planitia
Phobos 1 (USSR)	7 July 1988	attempted Mars orbiter and Phobos landers
Phobos 2 (USSR)	12 July 1988	Mars orbiter, some imaging before failure; Phobos landers failed

Table 2.1 continued

Mission	Launch date	Remarks
Mars Observer (USA)	25 Sept 1992	orbiter; contact lost during Mars orbit entry
Mars Global Surveyor (USA)	7 Nov 1996	orbiter, arrived 12 Sept 1997
Mars 96 (Russia)	16 Nov 1996	attempted orbiter/landers; launch failure
Mars Pathfinder (USA)	4 Dec 1996	lander/rover; landed 4 July 1997 in Ares Vallis
Planet-B, Nozomi (Japan)	4 July 1998	orbiter, atmospheric probe; arrival delayed to Dec 2003
Mars Climate Observer (USA)	11 Dec 1998	orbiter, lost on arrival at Mars 23 Sept 1999
Mars Polar Lander/ Deep Space 2 (USA)	3 Jan 1999	lander/descent probes; lost on arrival 3 Dec 1999
Mars Odyssey (USA)	7 Apr 2001	orbiter; currently conducting prime mission of science mapping
Mars Express/Beagle 2 (European Space Agency)	2 June 2003	orbiter/lander; arrival Dec 2003, prime aim is search for water and life from orbit and the surface
Mars Exploration Rovers (USA)	MER-A 10 June 2003 MER-B 25 June 2003	two identical rover missions (MER-A Spirit and MER-B Opportunity) to separate landing sites. Arrival Jan 2004, prime aim is search for past water

The first successful space mission was the Mariner 4 fly-by in 1964 and there followed various missions (both from the USA and the USSR) over the next 11 years, leading to the successful Viking 1 and 2 orbiters and landers.

2.2 The Mariner missions to Mars

Mariner 4 (launched 28 November 1964, fly-by of Mars 14 July 1965) was the first spacecraft to successfully reach Mars, flying as close as 9846 km. In total it returned 22 images of the planet from an on-board television camera, revealing a cratered, rust-coloured surface. The images were stored on a tape-recorder and took four days to transmit back to Earth (Figure 2.1).

■ How long will a radio signal travelling at the speed of light (3×10^8 m s^{-1}) take to travel between Mars and the Earth if the distance between the two planets at the time is 100 million kilometres, 100×10^6 km?

□ The time taken is the distance travelled divided by the speed. The distance 100×10^6 km is 100×10^9 m. Thus the time the radio signal will take to travel is:

$$\frac{100 \times 10^9 \text{ m}}{3 \times 10^8 \text{ m s}^{-1}} = 333 \text{ s}$$

which is about 5.5 minutes.

For help with powers of ten see Box 6.1 in the booklet Background material on maths skills.

Figure 2.1 One of the first images of Mars obtained by a spacecraft. This Mariner 4 image shows the crater named after it. The 151 km diameter Mariner crater is the large faint circle in the centre of the image. The image was taken from 12 600 km and covers 250 km by 254 km.

The delay in getting Mariner 4's images back to Earth was due to the slow rate at which the data could be transmitted, not the actual distance between the Earth and Mars. On average Mars is about 200 million kilometres from Earth. However, on 27 August 2003 the distance between Earth and Mars was the closest it has been for 60 000 years, a mere 55.76 million kilometres.

Mariners 6 and 7 (Mariner 6 launched 24 February 1969, fly-by of Mars 31 July 1969; Mariner 7 launched 27 March 1969, fly-by of Mars 5 August 1969) completed the first dual mission to Mars, flying over the equator and south polar regions and analysing the martian atmosphere and surface with remote sensors, as well as recording hundreds of images (Figure 2.2). By chance, both spacecraft flew over cratered regions of the planet so they did not observe the giant volcanoes that dominate the northern hemisphere of Mars nor the large canyon in the planet's equatorial region. The images returned covered about 20% of the planet and revealed a surface quite different from Earth's moon, somewhat contrary to the impressions left by

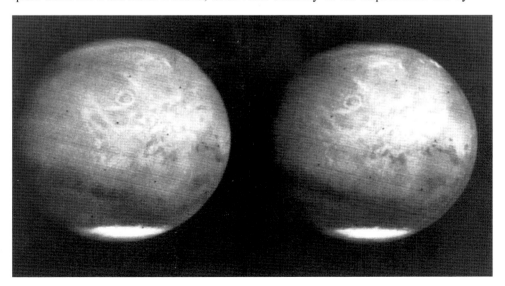

Figure 2.2 Mariner 7 view of Mars from 300 000 km. The Mariner 7 spacecraft and its twin (Mariner 6) were designed specifically to concentrate on Mars. Better quality imaging was planned to give a more complete picture of the martian surface to help in planning future missions to Mars to search for signs of life.

Mariner 4's images. Mariners 6 and 7 revealed cratered deserts, as well as depressions with no craters and huge concentrically terraced regions formed by impacts.

Mariner 9 was the final Mars mission in NASA's Mariner series of the 1960s and early 1970s. It was designed to be the first Mars orbiter, marking a transition in the exploration of Mars from fly-bys of the planet to spending time in orbit around it. Mariner 9 was launched successfully on 30 May 1971, and became the first artificial satellite of Mars when it arrived and went into orbit on 13 November 1971, where it functioned for almost a year. However, when it arrived, the planet was shrouded in an enormous dust storm and ground controllers had to wait until the surface was clearly visible so that Mariner 9 could begin compiling its global mosaic of high-quality images of the martian surface. The storm persisted for a month, but after the dust cleared Mariner 9 proceeded to reveal a very different planet to that expected from the fly-by images of Mariners 4, 6 and 7, one that boasted gigantic volcanoes and a huge canyon stretching 4 800 kilometres across its surface. More surprisingly, the relics of ancient riverbeds were carved in the landscape of this seemingly dry and dusty planet. Mariner 9 exceeded all of its initial photographic requirements by photo-mapping 100 percent of the planet's surface and providing the first close-up pictures of the two small, irregular martian moons: Phobos and Deimos.

2.3 The Viking missions

Viking 1, launched 20 August 1975, and Viking 2, launched 9 September 1975, had the primary aim of looking for evidence of life, past or present, on the surface of Mars. Each spacecraft consisted of an orbiter and a lander, Viking 1 entering orbit around Mars on 19 June 1976 and Viking 2 on 7 August 1976 (Figure 2.3).

Figure 2.3 The surface of Mars taken from the Viking 2 lander, which landed on Utopia Planitia on 3 September 1976.

Figure 2.4 A 360° panorama of the Mars Pathfinder landing site showing part of the Pathfinder spacecraft and the Sojourner rover investigating one of the nearby boulders.

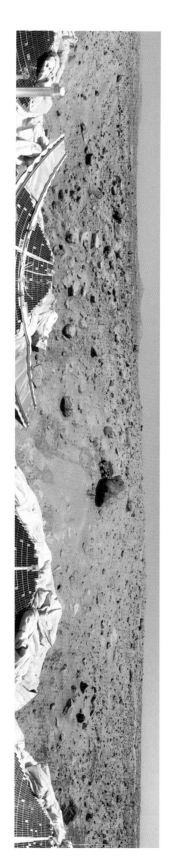

On arrival, the orbiters began taking pictures of the martian surface, from which landing sites were selected. Viking 1 soft-landed on the surface of Mars on 20 July 1976 and Viking 2 on 3 September 1976. The orbiters continued imaging and, between them, imaged the entire planet at what was then high resolution (Box 2.1). The orbiters also conducted atmospheric water vapour measurements and mapped parts of the planet using infrared radiation. The Viking 1 orbiter flew within 90 km of Phobos. The two Viking landers took full 360-degree pictures, collected and analyzed samples of the martian soil, and monitored the temperature, wind direction, and wind speed. The Viking missions revealed further details of volcanoes, lava plains, huge canyons, and the effects of wind and water. Analysis of the soils at the landing sites was the first attempt to look for life on Mars. We'll look at the lander biology experiments in more detail in Chapter 5.

2.4 The Mars Pathfinder Mission

The Mars Pathfinder mission (launched 4 December 1996, landed on Mars 4 July 1997) was only the third spacecraft to successfully land on Mars. The mission was unusual in that it had the primary objective of demonstrating the feasibility of low-cost landings on and exploration of the martian surface (Figure 2.4). The scientific objectives of the mission included atmospheric entry science, long-range and close-up surface imaging, and the general objective of being able to characterize the martian environment for further exploration. The spacecraft entered the martian atmosphere without going into orbit around the planet and landed on Mars with the aid of parachutes, rockets and airbags, taking atmospheric measurements on the way down. Pathfinder carried a small rover called Sojourner, which was a six-wheeled vehicle that was controlled by an operator back on Earth (Figure 2.5). The lander and rover operated until communication was lost for unknown reasons on 27 September 1997.

Figure 2.5 The Sojourner rover on the surface of Mars. Sojourner was a small rover with a mass of only 11 kg; its height was about 50 cm. The image was taken from the Mars pathfinder lander.

2.5 Mars Global Surveyor

Mars Global Surveyor was launched 7 November 1996, and entered orbit around Mars on 12 September 1997. After a year and a half trimming its orbit from a looping ellipse to a circular track around the planet, the spacecraft began its prime mapping mission in March 1999. It has observed the planet from a low-altitude, nearly polar orbit over the course of one complete martian year, the equivalent of nearly two Earth years.

One of the great improvements offered by the instruments on the Mars Global Surveyor was the **resolution** (see Box 2.1) of its camera, the Mars Orbiter Camera (MOC). In fact, the smallest feature detectable by this instrument on Mars' surface was of only 1.4 m in length. Mars Global Surveyor's mission was to map the entire planet at high resolution using the MOC, and gather data on the surface morphology, topography, gravity, weather and climate, surface and atmospheric composition, and planetary magnetic field. It will also provide relay for future Mars missions (i.e. it will act as a communications satellite in Mars orbit to relay data back to Earth from other missions). The mission has studied the entire martian surface, atmosphere, and interior, and has returned more data about Mars than all other Mars missions combined. Among its key science findings so far, Global Surveyor has taken pictures of gullies and debris flow features that suggest there may be current sources of liquid water on Mars. MOC's findings were supported by data from another instrument on the spacecraft, the Mars Orbiter Laser Altimeter (MOLA). For 27 months – longer than a martian year – MOLA gauged the daily height of the polar ice-caps, meticulously recording how much frozen material accumulated in winter and eroded in summer in each hemisphere. MOLA data gave scientists their first detailed views of Mars' north polar ice-cap (Figure 2.6).

BOX 2.1 RESOLUTION OF SPACECRAFT IMAGES

Resolution is an optical term, referring to the most closely spaced objects that can be distinguished as two separate objects in the optical image. Low resolution (or coarse resolution) means that closely spaced objects cannot be distinguished, whereas high-resolution images reveal fine detail. In the case of astronomical images of the sky, resolution is conventionally expressed as fractions of a degree, but resolution on planetary surfaces is more usefully expressed in terms of true distance on the ground.

In a digital image, the detail that can be seen usually depends on the size of the picture elements or *pixels* of which the image is composed. The resolution is typically twice the width of the pixels. The highest resolution images of Mars obtained by the Mars Global Surveyor MOC have pixels that represent 1.4 m across, but some of the surface has been imaged with the lowest (i.e. worst) MOC resolution, namely pixels that represent 230 m.

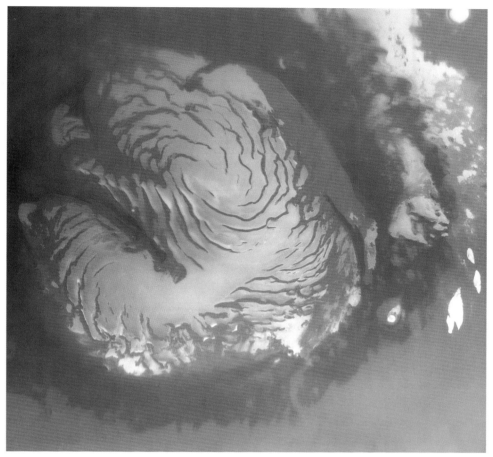

Figure 2.6 This view of the martian north polar ice-cap was acquired on 13 March 1999, near the start of the mapping phase of the Mars Global Surveyor mission. The light-toned surfaces are residual water ice that remains through the summer season. The nearly circular band of dark material surrounding the ice-cap consists mainly of sand dunes formed and shaped by wind. The north polar ice-cap is roughly 1100 kilometres across.

QUESTION 2.1

An image of Mars is produced by an orbiting spacecraft camera employing 512 pixels from top to bottom of the image and 1024 pixels from side to side. The area imaged corresponds to a scale of 4.5 km from top to bottom and 12.7 km from side to side respectively on the surface of Mars.

(a) Would it be possible to distinguish (or resolve) (i) 500 m-scale impact craters? (ii) 1 m-scale boulders?

(b) What happens (qualitatively) to the resolution if the same imaging system is used from a higher orbit? (Hint: consider how the size of the area that is imaged changes as the orbital distance increases.)

2.6 Mars Odyssey

Mars Odyssey was launched on 7 April 2001. It arrived at Mars on 24 October 2001. Using a process called aerobraking to slow down instead of firing thrusters, the spacecraft skimmed the surface of the martian atmosphere 332 times over nearly three months. The procedure saved more than 200 kilograms of fuel, which enabled NASA to launch it on a smaller and less expensive launch vehicle.

Mars Odyssey is targeted primarily to study the composition of the surface of Mars and to detect water and shallow buried ice. It also collects data on the radiation environment to help assess potential risks to any future human exploration and can act as a communications relay for future Mars landers. Its high-gain antenna unfurled on 6 February 2002, and its instruments began mapping Mars at the end of that month. Odyssey has several innovative instruments that obtain images of Mars in different parts of the electromagnetic spectrum (Box 2.2):

- THEMIS. The Thermal Emission Imaging System is a camera that obtains images of Mars in the visible and infrared parts of the spectrum in order to determine the distribution of minerals on the surface of Mars.
- GRS. The Gamma Ray Spectrometer uses the gamma-ray part of the spectrum to look for the presence of 20 elements (e.g. carbon, silicon, iron, magnesium). This has enabled scientists to produce maps of the distribution of these elements on the martian surface, for example Figure 2.7 shows the distribution of iron – the element (when combined with oxygen as iron oxide) that is responsible for the red colour of Mars.
- MARIE. The Martian Radiation Experiment is designed to measure the radiation environment of Mars.

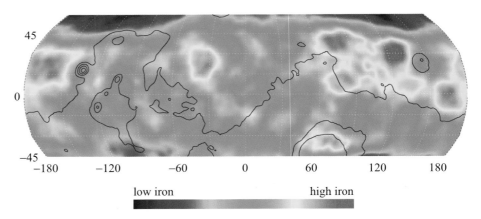

Figure 2.7 A map of the distribution of the element iron on the surface of Mars obtained by Mars Odyssey's gamma ray spectrometer. Areas with high amounts of iron are shown in red; those with low amounts of iron are shown in blue. The black lines are contours representing areas of the same elevation on the martian surface.

BOX 2.2 THE ELECTROMAGNETIC SPECTRUM

Visible light spans a range of wavelengths, from approximately 400 nm to 700 nm. Electromagnetic waves with wavelengths outside this range cannot, by definition, represent visible light of any colour. However, such waves do provide a useful model of many well-known phenomena that are more or less similar to light. For example, everyone is familiar with radio waves; we all rely on them to deliver radio and TV programmes. Radio waves are known to have wavelengths of about 3 cm or more; their well-established properties include the ability to be reflected by smooth, metal surfaces and to travel through a vacuum at the same speed as light.

One nanometre (nm) is 1×10^{-9} m.

The wide range of phenomena that can be modelled by electromagnetic waves is illustrated in Figure 2.8. As you can see, the full **electromagnetic spectrum**, as it is called, ranges from long wavelength **radio waves**, through **microwaves** and **infrared radiation**, across the various colours of **visible light** and on to short wavelength **ultraviolet (UV) radiation**, **X-rays** and **gamma-rays**. These various kinds of **electromagnetic radiation** arise in a wide range of contexts (as illustrated) but fundamentally they differ from one another *only* in the wavelength (and thus frequency) of the electromagnetic waves used to model them.

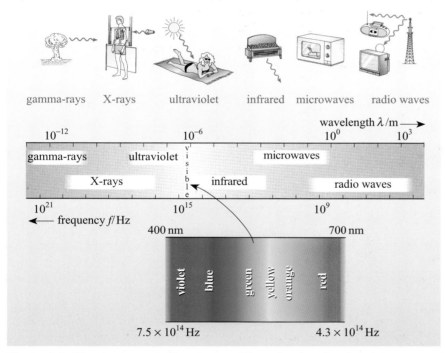

Figure 2.8 The electromagnetic spectrum. Note also that the ultraviolet (meaning 'beyond the violet') adjoins the visible violet, and the infrared (meaning 'below the red') adjoins the visible red. The boundaries of the various regions are deliberately vague; scientists and technologists often draw the divisions somewhat loosely.

2.7 The 2003–2004 missions

2.7.1 Mars Express/Beagle 2

The European Space Agency (ESA) launched the Mars Express spacecraft on 2 June 2003 (Figure 2.9). The orbiter will carry a remote observation payload with some re-use of European instruments lost on the ill-fated Russian Mars 96 mission, as well as a lander communications package to support Mars lander missions.

Figure 2.9 The launch of Mars Express by a Soyuz launch vehicle from Baikonur on 2 June 2003.

The mission's main goals are to search for subsurface water from orbit and drop the Beagle 2 lander on the martian surface. Beagle 2 is a lander designed specifically to study the martian surface environment and will be the first martian lander since Viking to carry instruments designed to detect evidence of present or past life.

Seven scientific instruments onboard the orbiting spacecraft will perform a series of remote sensing experiments designed to shed new light on the martian atmosphere, the planet's structure and geology.

Remote sensing refers to the obtaining and recording of information about a planet from a distance.

The orbiter's main objectives are:

- global high-resolution studies of martian geology at 10 m resolution
- global high-resolution mineralogical mapping of the martian surface at 10 m resolution
- global atmospheric circulation and high-resolution mapping of atmospheric composition
- studies of subsurface structures down to the permafrost, at km-scale
- studies of surface–atmosphere interaction
- studies of the interaction of the atmosphere with the interplanetary medium.

2.7.2 Mars Exploration rovers

NASA's Mars Exploration Rovers missions (named Spirit and Opportunity) were launched on 10 June and 7 July 2003 (Figure 2.10). Building on the success of the Mars Pathfinder rover, Sojourner, the 180 kg rovers are effectively mobile laboratories for the study of the martian surface. The mission's main scientific goals are to search for and characterize a wide range of rocks and soils that hold clues to past water activity on Mars. The spacecraft will be targeted to land at sites

Figure 2.10 An artist's impression of one of the Mars Exploration rovers on the surface of Mars.

One martian day is 24 hours and 37 minutes long.

that appear to have been affected by liquid water in the past. The rovers carry panoramic cameras as well as microscopic imaging equipment and tools for grinding away at rocks on the martian surface. Each rover is designed to carry out scientific investigations on the surface of Mars for up to 90 martian days.

2.7.3 Nozomi

Nozomi (known as Planet-B before launch) is the first Japanese Mars orbiter. It was launched on 4 July 1998, on the second flight of a new Japanese launch vehicle from Kagoshima Space Centre. The primary scientific objective of the Nozomi mission is to study the martian upper atmosphere with emphasis on its interaction with the solar wind (a continuous outward flow of electrically charged particles from the Sun).

2.8 Summary of Chapter 2

- The first successful mission to Mars was the US Mariner 4 mission launched in 1964.
- Mariner 9 was the first spacecraft to successfully enter orbit around Mars in November 1971.
- The twin Viking landers successfully soft-landed on Mars in 1976 undertaking the first scientific search for evidence of life on the planet.
- Mars Global Surveyor has undertaken the highest resolution mapping of the surface of Mars to date.
- Mars Odyssey carries a series of scientific instruments that are able to map the surface of Mars in different parts of the electromagnetic spectrum.

2.9 Question

QUESTION 2.2

Which *two* of the following statements about Mars spacecraft are *incorrect*?

A Mars Odyssey's gamma-ray spectrometer is able to detect electromagnetic radiation with a wavelength of 10^{21} m.

B Instruments on board the Mars Odyssey spacecraft would be able to detect electromagnetic radiation with a wavelength of 10^{-5} m.

C Instruments on board the Mars Global Surveyor spacecraft would be able to detect electromagnetic radiation with a wavelength of 6×10^{-7} m.

D Mars Odyssey's gamma-ray spectrometer is able to detect electromagnetic radiation with a wavelength of 10^{-12} m.

E Mariner 4 sent its data back to Earth using electromagnetic radiation with a wavelength of 10^{-3} m.

CHAPTER 3
INSIDE MARS

3.1 Introduction

We live on Earth, so it is only natural that we have a wealth of knowledge about the Earth's composition and structure. Much of this information is derived from the study of rocks, yet we do not know precisely what the Earth is composed of since direct geological evidence from rock samples tells us only about its outermost part. Determining the internal structure of other planets is even more difficult. In this chapter you will learn how we can use information about the basic physical properties of a planet to help us understand its interior and how theories on planetary formation help scientists understand how the different layers of a planet are produced.

At the present time, our knowledge about the interior of Mars is based on theoretical models of how a rocky planet like Mars was formed. This is supplemented by information on its composition, gravity and magnetic fields obtained by spacecraft and landers.

Much of this chapter is concerned with the evidence obtained by planetary scientists from studies of the composition and physical properties of the rocks and minerals that make up planets like Earth and Mars.

- From the information you met in Chapter 1, what is the fundamental difference in composition between the planets of the inner and outer Solar System?

- You saw in Chapter 1 that information from the Voyager and Galileo missions tells us that the giant planets of the outer Solar System, Jupiter, Saturn, Uranus and Neptune, are composed largely of gases. In contrast, the planets of the inner Solar System, Mercury, Venus, Earth and Mars, are rocky.

But what is this rock? A definition of what a rock is and how it differs from a mineral is given in Box 3.1. The rocks that make up the planets of the inner Solar System are composed mainly of the elements silicon and oxygen, which, in combination with various metallic elements such as iron and magnesium, make up a group of minerals known as **silicates**. We'll start our investigation of the interior of Mars by looking at the information we can obtain about the internal structure of planets from the basic physical properties of their rocks and minerals.

BOX 3.1 ROCKS AND MINERALS

A rock is a solid assembly of mineral grains. A mineral is a solid material that has formed by natural processes and has a chemical composition that falls within certain narrow limits. Its constituent atoms are arranged in a regular three-dimensional pattern and this determines the characteristic shape of the crystals that it forms. A rock may consist of just one type of mineral, but more usually it contains several different minerals.

A micrometre, μm, is one-millionth of a metre. $10^6\,\mu m = 1\,m$.

An atom is the smallest particle of a chemical element. It consists of a nucleus (containing protons and typically also neutrons) surrounded by electrons.

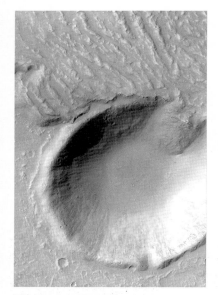

Figure 3.1 Extrusive igneous rocks on Mars. This Mars Global Surveyor image obtained in April 2003 shows the margin of a large lava flow on Mars. Some of the lava broke out and poured into an adjacent crater that was formed by meteor impact. The picture covers an area about 3 km wide.

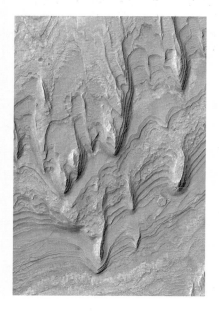

More than 3500 different minerals are known, but the number of common rock-forming minerals is much smaller. Mineral grains can be intact crystals or fragments, and can vary in size from a few micrometres to a few centimetres.

Different types of rocks form in different ways, and the processes of formation and any subsequent activity leave their marks on the rocks. The rocks that make up a planet's surface are manifestations of the geological activity of the planet. There are three main processes of rock formation, each of which produces characteristic features in the resulting rock.

Rocks that have solidified from a molten state are called **igneous** rocks. Heating within a planet's interior to temperatures around 1000 °C produces **magma** (which is molten rock). Magma may emerge onto the planet's surface via a volcanic eruption, in which case it is known as lava. The rock thus formed is called an **extrusive** igneous rock (Figure 3.1). Alternatively, magma may cool slowly while still underground forming so-called **intrusive** igneous rocks. As the magma cools, crystals grow from the liquid. The rocks that form underground may eventually be uncovered as overlying rocks become worn away through erosion of the surface. Igneous rocks are characterized by the presence of complete crystals and their size indicates the rate at which the magma cooled: in general, slow cooling produces large crystals.

Sedimentary rocks are formed by the deposition of layers of sediments (Figure 3.2). Sediments can originate from rocks that have been broken up by **weathering** (exposure to rain, wind and frost) and **erosion**. The resulting small fragments are then transported by water, wind or ice and deposited elsewhere in roughly horizontal layers known as **strata**. The grains of sedimentary rocks are usually fragments rather than complete crystals. When first deposited, sediments can be quite soft and malleable. However, over time, as they are gradually buried and compressed, they form a harder sedimentary rock.

The third main group of rocks are the **metamorphic** (meaning 'changed form') rocks. A metamorphic rock is formed when any type of rock is heated to temperatures of several hundred degrees Celsius and/or is subjected to high *pressure* because of the *weight* of overlying rocks. Unlike igneous rocks, metamorphic rocks do not cool from a liquid; rather, the change occurs while the rock remains in the *solid* state.

During metamorphosis, the atoms in the minerals making up the rock become reorganized, sometimes resulting in the formation of new minerals and changing the rock's appearance, for example pronounced bands may be produced, while maintaining the same chemical composition.

Figure 3.2 Sedimentary rocks on Mars. This Mars Global Surveyor image shows some of the eroded remains of the sedimentary rock that once filled an impact crater on Mars. The layers form terraces; wind has eroded the material to form the tapered, pointed ridges seen here.

Some common minerals and rocks

An important group of minerals in surface rocks of both Earth and Mars are the **carbonates**, which contain carbon and oxygen combined with elements such as calcium or magnesium. The mineral calcite is calcium carbonate and on Earth the sedimentary rock limestone is composed largely of calcite. Another group of minerals present on Earth and Mars are oxides, for example, iron oxide (rust) in which iron is combined with oxygen.

■ What significant aspect of the martian surface is iron oxide responsible for?

❏ Its colour. Iron oxide is red.

Common rock types

Different rock types give off different amounts of heat after absorbing sunlight all day. Data from the Mars Global Surveyor spacecraft (Section 2.5) for the amount of heat given off by martian rocks indicates that **basalt** or rocks of basaltic composition are the most common rocks on the surface of Mars (Figure 3.3). Basalt is an extrusive igneous rock, formed by rapid cooling and crystallization, which results in it having small crystals. It is very common on Earth, the Moon, and Mars.

On Earth, another common igneous rock is **granite**, which is an intrusive igneous rock that has cooled slowly underground. It contains relatively small amounts of calcium, iron or magnesium, but quite large proportions of silicon, sodium and potassium. Its slow cooling ensures that the mineral grains are large, giving the rock a coarse texture.

Common sedimentary rocks on Earth include **sandstones**, **mudstones**, **breccias** and **limestones**. Chemically, their composition varies according to that of the original rock. Sandstones are composed mainly of mineral grains, especially the mineral quartz, which is the natural form of silicon dioxide (silica). Mudstones contain clay minerals (which are also silicates). Breccias are rocks with large, angular rock fragments held together by finer-grained material.

Figure 3.3 A large boulder at the Mars Pathfinder landing site (the boulder was nicknamed Yogi by the mission scientists). The Sojourner rover, adjacent to the boulder, carried an instrument capable of providing basic chemical information about rocks. In the case of Yogi, it suggested that the rock was an extrusive igneous rock with a composition similar to common basalts found on Earth. The Sojourner rover is about 50 cm high.

There is strong evidence for the presence of sedimentary rocks in some of the images from the Mars Global Surveyor spacecraft (Figure 3.2), as well as from general interpretation of the formations on the martian landscape that resemble river channels and lake beds seen on Earth (see Section 4.4.3). However, exactly what kinds of sedimentary rocks there might be on Mars (e.g. sandstones or mudstones) is less certain. Breccias can be formed from the debris thrown out from impact craters. They are a common rock type found on the Moon, so they are a possible rock type to be found on the martian surface.

Slate and **marble** are examples of metamorphic rocks found on Earth. Slate originates from mudstones that have been heated to between 200 °C and 350 °C at depths of 5–10 km below the Earth's surface. Heating at greater temperatures and pressures produces a rock called gneiss, which has much coarser crystalline grains than slate but whose chemical composition still reflects that of the silicates of the original mudstone. Marble originates from limestone and generally contains only one mineral, calcite. As yet, there is no conclusive evidence for any types of metamorphic rocks on Mars.

3.2 Some basic physical properties of the planets

We can obtain a surprising amount of information about a planet by looking at some of its basic physical properties (Table 3.1). In this section we'll look more closely at one important physical property, density, and how it varies from planet to planet. Density can also be used to tell us about the composition of a planet.

Density is a measure of the mass per unit volume of a substance, and is defined by the equation:

$$\text{density} = \frac{\text{mass}}{\text{volume}} \tag{3.1}$$

SI units are an internationally agreed system of units used in science.

In SI units, mass is measured in kilograms and volume is measured in cubic metres, so the SI units of density are kilograms per cubic metre (kg/m^3 or kg m^{-3}). If you are unsure how to manipulate units, see Box 3.2. Density values of common materials can cover quite a wide range. Water has a density of 1.0×10^3 kg m^{-3}, whereas a rock such as granite is around 2.7×10^3 kg m^{-3}, and iron is 7.9×10^3 kg m^{-3}. In other words, a cubic metre of granite would weigh 2700 kg, or 2.7 tonnes! Since one cubic metre is somewhat larger than, for instance, the average pebble or rock you might pick up on a beach, these large numerical values for density are often difficult to grasp. Instead, it is often convenient to think of densities in smaller units, so you may come across, or prefer to think of, density values expressed as grams per cubic centimetre (i.e. g cm^{-3}). Thus a density of 2.7×10^3 kg m^{-3} could be expressed as 2.7 g cm^{-3}. However, when making calculations involving density, always ensure you use the SI units for mass (kg), volume (m^3) and density (kg m^{-3}).

There are 1000 kg in 1 tonne.

BOX 3.2 MANIPULATING UNITS

You can combine two measured quantities by multiplying or dividing one by the other. This produces a quantity with new units. Units can be multiplied or divided just like any other symbols.

For example, if a wall is 3.2 m high and 4.5 m wide. What is its area?

To find its area, multiply the height and width together:

$$\text{area} = 3.2 \text{ m} \times 4.5 \text{ m} = 14.4 \text{ m}^2$$

The SI units of area are metres $\times$ metres, that is metres squared, which is usually expressed as square metres (m^2).

If we want to calculate the density of a planet we can do this if we know its mass and volume. For example, the Earth has a mass of 5.97×10^{24} kg and a volume of 1.083×10^{21} m^3. What is its density?

To find the density, divide the mass of the planet by its volume:

$$\text{density} = \frac{5.97 \times 10^{24} \text{ kg}}{1.083 \times 10^{21} \text{ m}^3} = 5.51 \times 10^3 \text{ kg m}^{-3}$$

The answer has SI units of $\dfrac{\text{kilograms}}{\text{cubic metres}}$, abbreviated to kg/m^3 or kg m^{-3}.

Table 3.1 Basic physical data for the planets.

	Mercury	Venus	Earth	Moon[a]	Mars	Jupiter	Saturn	Uranus	Neptune	Pluto
mass										
/10^{24} kg	0.330	4.87	5.97	0.074	0.642	1900	569	86.8	102	0.013
/Earth mass	0.055	0.815	1.00	0.012	0.107	318	95.2	14.4	17.1	0.002
mean distance from Sun										
/10^6 km	57.91	108.2	149.6	149.6	227.9	778.4	1427	2871	4498	5906
/AU	0.39	0.72	1.00	1.00	1.52	5.20	9.54	19.19	30.07	39.48
orbital period[b]	88.0 days	224.7 days	365.0 days	27.3 days	686.5 days	11.86 years	29.42 years	83.75 years	163.7 years	248.0 years
axial rotation period[c] /days	58.6	243	0.997	27.3	1.03	0.412	0.444	0.718	0.671	6.39
mean radius/km	2440	6052	6371	1738	3390	69 910	58 230	25 360	24 620	1137
density/10^3 kg m^{-3}	5.43	5.20	5.51	3.34	3.93	1.33	0.69	1.32	1.64	2.1
mean surface temperature/°C	170	500	15	−23	−50	−	−	−	−	≥ −233
number of satellites	0	0	1	0	2	≥ 61	≥ 31	≥ 21	≥ 8	1

[a] The Moon, although strictly not a planet, is included for comparison.

[b] Orbital period is the time the planet takes to complete one revolution, i.e. orbit, around the Sun (for the Moon it is the time for one orbit around the Earth).

[c] The axial rotation period is the time the planet takes to complete one rotation on its axis (i.e. equivalent to 1 day for the Earth).

One means of learning something about the interior of a planet is to look at its density relative to its size. We can do this by constructing a graph that shows how planetary radius varies with density, using the quantities given in Table 3.1. This sounds reasonably straightforward, but looking at the range of radii you have to consider, perhaps it is more complicated.

■ Why would it be difficult to plot the radii of planetary objects on a linear graph?

☐ The range in radius values is so large (from a few 1000 km to 70 000 km) that the data would not all usefully fit on an ordinary linear graph since the smaller-sized objects would all be crammed together at one end.

This problem is overcome by using a special type of graph where one of the axes increases in intervals of a power of ten, for example 10, 100, 1000, 10 000 and so on. We can use that axis for the radii of the planets and an 'ordinary' linear axis, for their densities, as shown in Figure 3.4. Such graphs are a convenient way of plotting data that range across many powers of ten. The axis in which an increase or decrease of one unit represents a tenfold increase or decrease in the quantity measured is called a logarithmic axis.

Looking at Figure 3.4, four planets form an obvious family, characterized by their huge dimensions (radii greater than 10 000 km) and low densities. These are the giant planets, which are dominated by Jupiter, the mass of which is more than that of all the other planets put together.

A second group straggles off towards the high-density area at the right-hand side of the graph. This group includes the terrestrial planets Mercury, Venus, Earth and

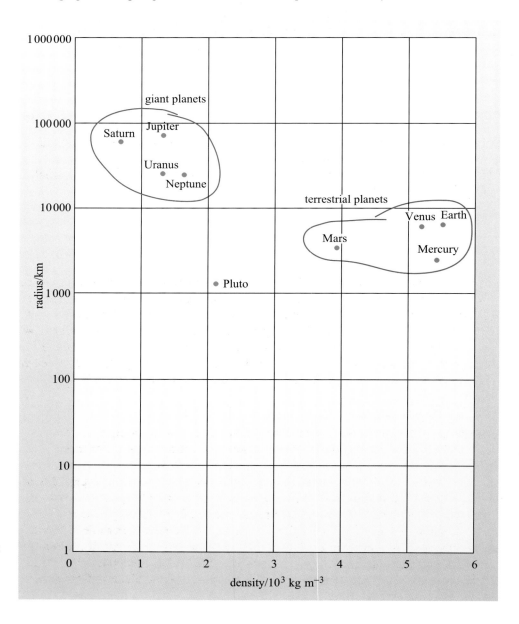

Figure 3.4 A graph of the radii of the planets in kilometres plotted on a logarithmic axis against the densities of the planets in kilograms per cubic metre.

Mars, which have densities ranging from 3.9×10^3 kg m^{-3} to 5.5×10^3 kg m^{-3} and mean radii between 1000 km and 10 000 km. Their densities suggest that these bodies are predominantly rocky, like the Earth, which is why they are called the terrestrial planets.

- ■ The minerals that make up rocks have densities between 2.5×10^3 kg m^{-3} and 3.5×10^3 kg m^{-3}. Is it likely that the terrestrial planets could consist *exclusively* of rocky material?

- ❏ No, they must contain a denser component as their densities (particularly the densities of Earth, Venus and Mercury) are considerably higher.

It is important to appreciate that the calculation of mass/volume gives us a mean value for the density of the planet. Planets can be made of layers of material that have quite different densities. For example a planet may have high-density material (such as iron) at its core and somewhat lower-density material (such as rock) nearer the surface.

> The mean value is the average of a set of quantities together. It is found by adding a group of quantities and dividing by the number of quantities in the group.

QUESTION 3.1

Suppose a planet is composed of two layers, an inner core composed of iron and an outer layer composed of a silicate rock. If there are equal quantities of iron (density = 7.9×10^3 kg m^{-3}) and a silicate rock (density = 2.7×10^3 kg m^{-3}), what would be the planet's mean density?

3.3 How to make a planet

In Chapter 1, you saw the incredible diversity that makes up the planets of the Solar System. In this section, we'll look at our present understanding of how the planets formed. We'll be paying particular attention to the processes that led to the formation of the rocky planets of the inner Solar System since the theoretical models for their formation help us understand the internal structure of Mars. We'll also use these theoretical models in Chapter 4 to help us understand some of the geological processes that have given rise to some of the dramatic features of the martian landscape.

3.3.1 Theories on the origin of the Solar System

Theories as to how the Solar System formed are by no means complete, with details of some particular processes being poorly understood. But the overall formation scenario that is most widely accepted was first put forward in its original form in 1796 by the French scientist Pierre-Simon Laplace. Laplace suggested that the Solar System was formed by the collapse of a large, initially spherical, rotating cloud of gas and dust. The term *dust* in this context refers to solid particles or 'grains' that have formed by the agglomeration of individual atoms, molecules, or smaller dust grains. They can be thought of as fragments of rock and ice, often with rather random structures. Modern work has refined this idea, but the overall theme remains. The processes that led to the formation of the planets of the Solar System are summarized below and in Figure 3.5.

> Agglomeration is the process of gathering things together to form a larger mass.

Figure 3.5 The processes that led to the formation of the planets of our Solar System. (See text for descriptions.)

CHAPTER 3 INSIDE MARS

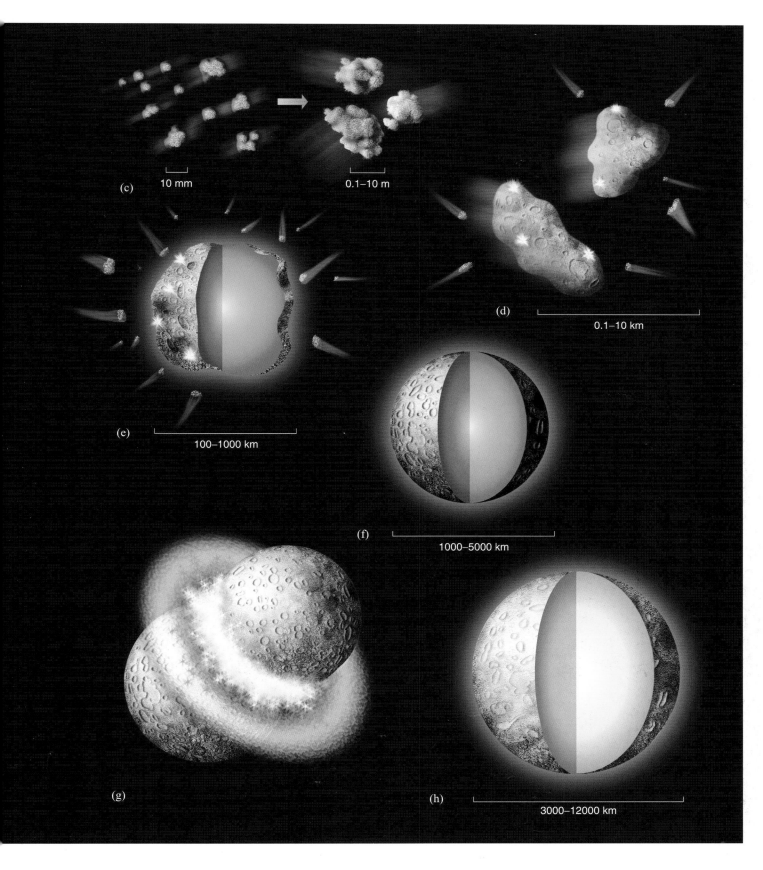

(c) 10 mm → 0.1–10 m
(d) 0.1–10 km
(e) 100–1000 km
(f) 1000–5000 km
(g)
(h) 3000–12000 km

45

The solar nebula

The formation process started with an enormous tenuous cloud of gas and dust, which underwent contraction due to mutual gravitational attraction (Box 1.1). The contracting cloud would also have had some rotational motion and, as it contracted further under gravity, the rate of spin would have increased, rather like the way a spinning ice skater rotates faster when they pull in their arms. As the resulting cloud spun, it flattened into a disc, with the young Sun at the centre (Figure 3.5a). This structure is referred to as a **protoplanetary disc** or **solar nebula** (Figure 3.5b). Such discs can be observed around some other stars, as shown in Figure 3.6.

Any two bodies that have mass will experience a force of gravitational attraction.

■ What is one consequence of the rotation of the protoplanetary disc for the orbits of the planets in our Solar System?

❏ It explains why all the planets orbit the Sun in the same direction.

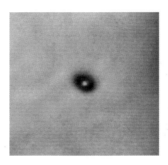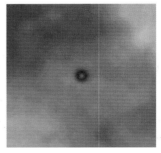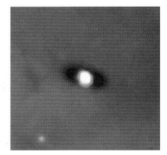

Figure 3.6 Images showing young stars with protoplanetary discs around them, taken with the Hubble Space Telescope. These discs will probably result in solar systems.

The process of accretion

Within the protoplanetary disc, the dust particles collided and stuck together. As more and more stuck together, larger particles were formed (Figure 3.5c), a process known as **accretion**. Over a period of about 10 000 years, these random collisions of dust and particles built up into clumps up to 10 mm across. These clumps of matter would have been of silicate composition in the inner region of the Solar System. The accretion process continued, eventually building clumps of 0.1–10 m in size.

> It is important to consider the temperature distribution of the protoplanetary disc: the material in the disc was hotter near the young Sun; the temperature decreased towards the edges of the disc.

By splitting sunlight into a spectrum it is possible to accurately determine the chemical composition of the Sun.

Scientists have a good idea as to what the chemical composition of the solar nebula was largely by studying the present day chemical composition of the Sun, since the outer part of the Sun has not changed its composition through time. Such studies tell us that the solar nebula was composed of about 75% hydrogen, 23% helium, 1% oxygen and about 1% of all the other elements. One consequence of the temperature distribution within the protoplanetary disc is that it would affect the kind of materials that would form as it cooled.

As the disc cooled, different materials would condense out in much the same way as water condenses from clouds of cooling steam.

- Will all the materials in the solar nebula condense at the same temperature?

- No, '**volatile**' materials (i.e. those with low boiling points) such as hydrogen, helium and other substances such as water, methane, ammonia and nitrogen will condense at lower temperatures than less 'volatile' materials such as silicates.

- How might the change in temperature within the protoplanetary disc, from the inner hotter regions (temperatures greater than 200 °C) to the outer colder regions (temperatures less than 0 °C), affect the kind of materials that might form?

- In the inner, hotter regions, materials such as small ice particles could not exist and so accretion of rocky and metallic material would dominate. In the outer regions, however, volatile materials would remain frozen and could eventually form planets.

The path to an embryonic terrestrial planet

After about 100 000 years, further random collisions during the accretion process had produced a profusion of bodies of 0.1–10 km across (Figure 3.5d). These objects, which were tiny planets called **planetesimals**, were the 'building blocks' of the planets.

Above diameters of about 10 km, the influence of the gravity from the larger planetesimals would have resulted in increasingly frequent collisions (Figure 3.5e).

- What effect would the gravitational attraction of these larger planetesimals have on the smaller ones?

- Nearby smaller planetesimals would be 'swept up' by larger bodies.

Where the growth of one planetesimal outpaced that of its neighbouring rivals, it may have developed into a **planetary embryo**, which is the term given to accreted bodies reaching a few thousand kilometres in diameter (Figure 3.5f). It is estimated that within a few thousand years the growth of these planetary embryos would have captured and incorporated most of the smaller planetesimals and, as a result, perhaps only a few hundred large planetary embryos would have replaced the initial profusion of planetesimals.

Having exhausted the nearby supply of planetesimals, the era of frequent collisions was over. The next stage of growth would have been slower and characterized by chance collisions between planetary embryos (Figure 3.5g). Giant impacts between embryos could have resulted in fragmentation of both, with the debris then recombining into a single mass. Moreover, the heat released during such massive impacts was, in some cases, capable of melting the newly combined mass, creating a body with a molten surface termed a 'magma ocean'. As the amalgamated body began to cool a thin crust would have developed on its surface (Figure 3.5h). This crust would have been continually destroyed and reformed as impact debris fell back to the surface.

Melting caused by impacts is known as impact melting.

Assembling a terrestrial planet

These devastating collisions between planetary embryos, and their associated impact melting would have allowed denser materials to segregate inwards and lighter materials to work their way outwards, leading to the crude separation into a core composed mainly of nickel and iron and an outer mantle composed mainly of rocky silicate material (Figure 3.5h). This process is known as **differentiation** and is explained in more detail in Box 3.3.

> The core of a planet is the densest, central part, differing markedly in composition from the layer above it.

Giant impacts would have continued to occur between these larger, partially differentiated bodies, resulting in the amalgamation of denser core material and the assembly of even larger bodies. It is estimated that the planets Mercury, Venus, Earth and Mars would have taken about 10 million years to reach half their current mass, and about 100 million years to fully complete their growth and build the planets through a process of chance collisions between planetary embryos. You should not assume that planetary bodies developed by the consecutive addition of planetary embryos of one particular size. It is much more probable that embryo collision first produced a number of larger embryos of varying sizes. These then collided to create even more massive bodies. Eventually a few large amalgamated bodies, resulting from either a few or many such collisions, were assembled into the planets. Although we can perhaps never know the sequence of such collisions, and the relative sizes of the bodies involved, what is certain is that the resulting giant impacts would have been cataclysmic events.

QUESTION 3.2

Estimate how many planetary embryos of mass 5×10^{22} kg would have been required to assemble (a) Mars, and (b) Earth. The mass of each planet is available in Table 3.1.

The later evolution and further internal differentiation of these planets was from then on driven mainly by processes within each body. Once the last giant impact in the inner Solar System had occurred, there were just four surviving terrestrial planets (Mercury, Venus, Earth and Mars).

BOX 3.3 DIFFERENTIATION AND THE ORIGIN OF PLANETARY LAYERING

The process of separating out the different constituents that make up a planetary body as a consequence of their physical or chemical behaviour is known as differentiation. One means of separating constituents is by allowing them to begin to melt. When a rock is heated, different minerals within the rock will melt at different temperatures. This phenomenon is known as **partial melting** and is a key process in the formation of liquid rock or magma. Once the constituents have been mobilized in this manner, they will begin to migrate under the influence of pressure or gravity.

> Planetary body is a general term, encompassing planets, their satellites and minor bodies (asteroids and comets).

- Imagine a body the size of a planetary embryo that had accreted from nickel, iron and silicate minerals. A typical mixture of nickel–iron has a density of about 7.9×10^3 kg m^{-3}, and a melting point some hundreds of degrees higher than silicates. What would happen if temperatures within this planetary embryo were increased to a point at which the melting of silicates began?

❑ Since nickel–iron has a higher melting point it would remain solid after the silicates had begun to melt and, because it is much denser than any silicate minerals, it would begin to sink towards the centre of the body.

If temperatures continued to rise beyond the melting point of nickel–iron itself, then an emulsion of liquid silicate and liquid nickel–iron would form. However, separation would still occur because the globules of nickel–iron would not mix into the silicate and so would continue to sink under gravity towards the centre of the body. In effect, the nickel–iron would 'rain out' from the emulsion towards the centre of the body. This is known as the 'rain-out' model and is a process of differentiation that is thought to lead to the first stage of formation of a planet's core and the development of associated layering.

An emulsion is a mixture of two immiscible liquids, like oil and water, in which one substance is suspended in another.

■ What would happen if a fast-moving object such as a planetesimal collided with a larger one, such as a planetary embryo, during the accretion process?

❑ Whilst some of the energy would be retained by fragments flung away from the impact site, a considerable amount of the energy of the colliding bodies would be converted to heat.

It is likely that such melting processes operated during the final stages of terrestrial planetary growth because of the heat generated by the collision of planetary embryos. The energy released is dependent upon their relative velocities and masses. Consequently, the energy released from collisions between planetary embryos would, in many instances, be sufficient to cause a considerable degree of melting. In those instances where particularly large embryos were involved, it may have been enough to melt entirely their amalgamated masses. Consequently, the resulting combined body would be able to differentiate due to its molten state, and so begin to develop layering (Figure 3.5h).

According to the 'rain-out' model, it is estimated that once core formation had begun, it was an extremely rapid process taking only a few tens of thousands of years to reach completion. Such rapidity, however, has important implications regarding the timing of core formation. It is probable that long before the planets were fully assembled, some differentiation and separation of a nickel–iron component had already begun within the planetary embryos due to melting that resulted from earlier embryo–embryo collisions. However, the formation of the core of an assembled planet, such as Mars, could not have been fully completed until all of the iron delivered via impacting bodies had arrived. Therefore, the formation of planetary cores must have occurred simultaneously with the assembly of the planet. It would have begun at the point at which the larger planetary embryos became molten and started to differentiate, and continued until the stage where amalgamation of these embryos had assembled a planet in which the separation of all the delivered nickel–iron could reach its completion.

3.4 The internal structure of the Earth

We saw in the last section how the process of differentiation can lead to the development of layers of different composition within a planet.

- ■ How does the presence of layers explain why the mean density of a planet is higher than the density of the rocks at its surface?

- ❏ Different layers within a planet must have different densities. If the mean density is higher than the density of the surface rocks, then there must be material at depth with a greater density.

We know more about the internal structure of the Earth than any other planet. However, given that the planets Mercury, Venus, Earth and Mars are thought to be of a broadly similar bulk composition (i.e. silicates, iron and nickel), albeit in different proportions, the processes that formed their layers are likely to have been similar in each case.

Scientists often use the terms **core**, **mantle** and **crust** to denote the different layers that exist within our planet. These layers display distinct properties based on the various mineral and rock compositions within the Earth (Figure 3.7). It could be argued that the atmosphere and hydrosphere (i.e. the oceans and ice-caps) are also layers within our planet. The different layers within the Earth have been identified through a variety of methods and observations.

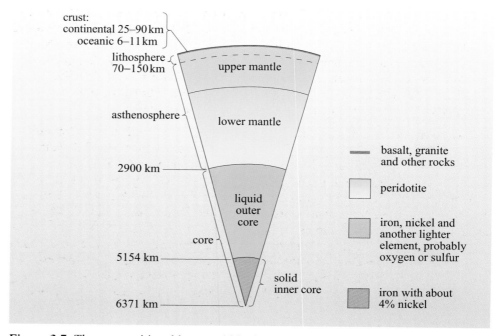

Figure 3.7 The compositional layers within the Earth. The lithosphere is the solid, upper layer of the Earth that includes the crust and the uppermost mantle. The asthenosphere is a relatively weak layer beneath the lithosphere that is deformable over long time periods. The thicknesses of the crust and lithosphere have been exaggerated for clarity.

3.4.1 Seismic evidence

The Earth has a radius of about 6400 km, yet even the deepest mines can only provide us with samples from a maximum depth of about 15 km, in other words only the outermost 0.23% of our planet! Consequently, most of the evidence regarding Earth's internal structure and composition comes not from rocks, but indirectly from **seismic waves** (from the Greek *seismos* meaning earthquake). These waves are vibrations or shocks that can pass through the Earth and be recorded by geophysical monitoring equipment. They can be created by either natural earthquakes or artificial explosions. Since these vibrations often pass through the deep interior of the Earth, monitoring them provides estimates of important physical properties, such as the speed of seismic waves deep within the planet. Seismic-wave speed and the manner in which the waves pass through the deeper levels are related to the mineralogical composition and hence the chemical composition of the material. If, for example, the Earth's composition and density were uniform, then the speed of propagation of seismic waves would be the same throughout the Earth's structure. However, experiments that monitor and record waves arriving at seismic stations generated by both artificial and natural seismic events demonstrate that this is not the case. In fact, distinct layers can be identified from the seismic data (Figure 3.7). The Earth's outermost 'skin' is known as the crust, which is 25–90 km thick in continental areas, and 6–11 km thick in the ocean basins. At the base of the crust is the mantle, which extends to a depth of nearly 3000 km. Seismic data indicate a boundary at 2900 km between the Earth's mantle and its core.

3.4.2 The core

There are no methods available for sampling or obtaining core material. However, using seismic and magnetic data, it is now believed that the *inner* part of the core is probably an alloy (i.e. a mixture) of iron with about 4% nickel. Iron is likely to be present since it is an abundant element in the Solar System, and the fact that it conducts electricity can help explain the origin of the Earth's magnetic field. Seismic evidence indicates that the *outer* core is liquid and broadly similar in composition to the inner core (i.e. predominantly iron and nickel).

Additional evidence indicating part of the core is liquid comes indirectly from the fact that the Earth has a magnetic field. In fact, the Earth's magnetic field is the strongest of all the terrestrial planets. Traces of ancient magnetism have been frozen into some of the Earth's earliest volcanic rocks and demonstrate that this type of magnetic field has been in existence for much of Earth's history. There is no known solid that has magnetic properties above 1200 °C and, since the lower mantle and core are known to be considerably hotter than this, the only way that the Earth's magnetic field could be generated and maintained is if the core conducts electricity and is in motion. It is the constant 'stirring up' of the liquid core due to the Earth's rotational motion (i.e. once every 24 hours) and the transfer of heat by convection currents within the liquid that generates Earth's strong magnetic field.

> Magnetism is a physical phenomenon produced by the motion of electric charge.

3.4.3 The mantle

Geological processes can bring samples of the upper part of the mantle to the Earth's surface. Lateral movement in our planet's outermost layers creates folding and buckling in crustal rocks, especially on the margins of the huge areas that make up the continents. Such areas, together with those forming the ocean floors, comprise huge 'plates; and the motion of these plates is known as **plate tectonics** (see Box 4.2). Plate tectonic processes, together with deep-seated volcanic activity, can bring rock material, which originally formed at greater depths, nearer to the surface. For instance rocks formed deep within the Earth can be broken off and carried to the surface in molten lava as exotic nodules or **xenoliths** (from the Greek *xenos* meaning strange or foreign, and *lithos* meaning rock) that can then be preserved inside lava flows when this molten rock solidifies at the Earth's surface. Alternatively, the buckling, folding and faulting of the Earth's crust, together with surface erosion, can eventually lead to the exhumation of materials that originated at a similarly deep level.

These rocks typically have the composition of **peridotite**, a rock type that is rich in the minerals olivine and pyroxene, which are less common minerals in the Earth's crust. Peridotite is also thought to constitute the mantle material of the other terrestrial planets.

> Pyroxene and olivine are important rock-forming silicate minerals.

3.4.4 The crust

The crust is the geologically highly active outer part of the Earth's layered structure and has two distinct compositions: the material that comprises the continents, and the material that forms the floor of the ocean basins. Silicate minerals dominate both of these crustal compositions, but the ocean crust in which iron and magnesium-rich minerals are more common, is more homogeneous, being largely composed of basalt. By contrast, the continental crust is much more variable in its composition because it is composed of many different rock types (metamorphic, sedimentary and igneous rocks) that have been added to the continental areas throughout geological time. However, continental crust may be considered to be broadly similar to granite in its composition. Iron- and magnesium-rich minerals are generally less common in the continental crust than in the ocean crust, so continental areas are the least dense part of the Earth's layers.

> It is important to realize that the difference in composition between the Earth's crust and mantle, although significant, is much less than that between the mantle and the nickel–iron core. In both mantle and crust the dominant minerals are silicates.

3.5 The internal structure of Mars

You've seen that most of the information regarding the nature of the deep layering of the Earth is derived from seismic studies, with additional chemical and mineralogical information about the topmost 100 km of mantle and crust coming from rock samples. In the case of Mars, however, we have no seismic data and little direct compositional information. However, important information about its internal structure and composition can be obtained from determining its density and looking at the variation in gravity across the planet using data collected by orbiting and fly-by spacecraft. Furthermore, by assuming that similar processes operated during the evolution of Mars and during the evolution of the Earth, and by comparing the mean density of Mars with the composition of surface materials determined by both orbiting and lander probes, scientists have deduced that Mars also has a layered structure.

In this section we'll look in more detail at the information we can obtain from using the density of Mars to investigate its internal structure and at some new information obtained by the Mars Global Surveyor spacecraft that is making scientists rethink what they know about the interior of the planet. We'll start by considering whether the Mars' density data indicate whether the rocks at deeper levels of the planet may be different from the rocky materials that are thought to dominate its surface. It is possible to calculate the density of Mars very accurately because both its size and mass can be determined.

- ■ The mass of Mars has been determined as 0.642×10^{24} kg. Calculate the mean density of Mars, which has a volume of 1.634×10^{20} m^3.
- ❏ We can obtain the average or mean density of Mars by dividing its mass by its volume (Equation 3.1). This gives us a density of 3.93×10^3 kg m^{-3}.

QUESTION 3.3

The most common rock believed to comprise the surface of Mars is basalt (Box 3.1), which has a density of 3.0×10^3 kg m^{-3}. How does this density compare with the mean density for Mars that you calculated above and what does this suggest about the variation of density with depth on Mars?

If you look at the data in Table 3.1 you can see that planets such as the Earth (mean density 5.51×10^3 kg m^{-3}) and Venus (mean density 5.2×10^3 kg m^{-3}) have much higher densities than Mars.

- ■ Look at the data in Table 3.1. Which planetary body in the inner Solar System has a density closest to that of Mars?
- ❏ The Moon, which has a density of 3.34×10^3 kg m^{-3}.

The Moon is the only other body in the Solar System for which we have seismic data, and these data were obtained from experiments carried on several of the Apollo missions. Consequently, we know quite a bit about its internal structure, in particular the likely maximum size of its core. As we don't yet have any seismic data for Mars, it is instructive to look at the Moon (it has a similar density to Mars) and see what we can infer about the internal structure of Mars. The Apollo seismic data and detailed mapping of the lunar surface suggest that the Moon's crust is on average about 70 km thick, exceeding 100 km in some of the highland regions and falling to about 20 km below some of the large impact-produced basins. The outermost layer of the Moon (equivalent to the Earth's crust and uppermost mantle) is rigid to a depth of about 1000 km and, like the Earth, is largely composed of peridotite. Estimates of the size of the Moon's core suggest it is between 220 km and 450 km in radius. The Moon has no magnetic field, suggesting that the core is solid.

The similarity in density between the Moon and Mars led scientists to believe that Mars' core must be relatively small; initial estimates put its size between 1300 km and 2000 km in radius. Mariner 4 also showed that Mars lacked a strong magnetic field. This was interpreted as indicating that Mars did not have a liquid core.

However, these initial interpretations have been dramatically changed by recent studies, which suggest that Mars may have a molten iron–nickel core, i.e. the interior of the planet is similar to the Earth. In March 2003, scientists from NASA's Jet Propulsion Laboratory finished analysing three years of radio tracking data from the Mars Global Surveyor spacecraft. They concluded that Mars has not cooled to a completely solid iron core, but that its interior is made up of either a completely liquid iron core or a liquid outer core with a solid inner core.

The scientists used the tracking of a radio signal emitted by the Mars Global Surveyor spacecraft to determine the precise orbit of the spacecraft around Mars. As with all the planets, Mars is influenced by the gravitational pull of the Sun. This causes a bulge towards and away from the Sun (similar in concept to the tides in Earth's oceans that are caused by the gravitational pull of the Moon). On Mars this bulge is extremely small, but it exerts a detectable force on the spacecraft. By measuring the effect of this force on the spacecraft's orbit, and using data from other Mars missions, the scientists were able to determine how flexible Mars is; this indicated that the core of Mars cannot be completely solid iron but must be at least partially liquid. In addition to the detection of a liquid core for Mars, the results also suggested that the size of the core is about one-half the size of the planet (Figure 3.8), as is the case for Earth and Venus.

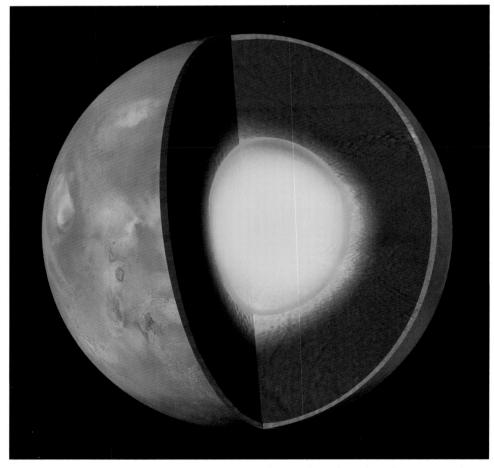

Figure 3.8 An artist's concept of the interior of Mars, based on data that suggest that Mars' core may be at least partially liquid and be about half the size of the planet.

3.6 Summary of Chapter 3

- Our knowledge of the internal structure of Mars is based on theoretical models of how rocky planets form, supplemented by information about its composition, gravity and magnetic fields obtained by spacecraft and landers.
- Density is a measure of the mass per unit volume of a substance, and is defined by the equation:

$$\text{density} = \frac{\text{mass}}{\text{volume}}$$

- The calculation of mass/volume gives us a mean value for the density of a planet. Planets can be made of layers of material that have quite different densities.
- In 1796, Pierre-Simon Laplace was the first scientist to suggest that the Solar System was formed by the collapse of a large, initially spherical, rotating cloud of dust and gas.
- The cloud of dust and gas contracted due to mutual gravitational attraction, flattening into a spinning disc with the young Sun at its centre, known as a protoplanetary disc.
- Within the protoplanetary disc, dust particles collided and stuck together gradually forming larger and larger particles, a process known as accretion.
- The material in the protoplanetary disc was hotter near the young Sun; the temperature fell towards the edges of the disc.
- Further accretion led to the formation of bodies 0.1–10 km across, known as planetesimals.
- Where the growth of one planetesimal outpaced that of its neighbouring rivals it may have developed into a body a few thousand kilometres across, which is known as a planetary embryo.
- Collisions between planetary embryos, and their associated impact melting, allowed denser material to segregate towards the centre of the embryos, a process known as differentiation.
- Detail of the internal structure of the Earth is provided by both direct and indirect evidence. The composition of the crust and upper mantle can be determined by examination of key rock types. The structure and composition of the deep mantle and core is revealed from seismic studies.
- Density data indicate that the rocks at deeper levels in Mars are more dense than the rocks at its surface.
- Mars has no significant magnetic field. However, data from the Mars Global Surveyor spacecraft suggest that it has a partially liquid core composed of iron and nickel.

3.7 Questions

QUESTION 3.4

Briefly summarize the main stages of planet formation that could have led to the formation of Mars.

QUESTION 3.5

Which *one* of the following statements correctly defines the planetary science term 'differentiation'?

A   The formation of planetary embryos.

B   Laplace's theory for the origin of the Solar System.

C   The process whereby the interior of a planetary body segregates into layers of different composition.

D   The process whereby large planetesimals attract smaller ones as a result of gravitational attraction.

QUESTION 3.6

Which *one* of the following is the correct and complete listing of the relative mean densities (in order of decreasing density) of the terrestrial planets?

A   Mars, Earth, Venus, Mercury

B   Earth, Mercury, Venus, Mars

C   Earth, Venus, Mars, Mercury

D   Mars, Venus, Mercury Earth

QUESTION 3.7

Which *two* of the following features does Mars share with Earth?

A   A mean density of $5.5 \times 10^3 \, \text{kg m}^{-3}$.

B   A strong magnetic field.

C   A partially liquid core.

D   A day length of just under 24 hours.

E   A mantle composed of peridotite.

QUESTION 3.8

Figure 3.9 shows a series of ripples or dunes on the surface of Mars in an area known as Candor Chasma. From your general knowledge of sand dunes on Earth, which *one* of the following statements is most likely to be an accurate inference about the rocks that make up the dunes?

A   The dunes and ripples are composed of large boulders of basalt and are therefore igneous rocks.

B   The dunes and ripples are composed of wind-blown fine sediment and are therefore sedimentary rocks.

C   The dunes are formed by a series of extrusive igneous rocks from a volcano.

D   The dunes were formed by an igneous intrusion.

E   The dunes and ripples are composed of loose particles of fine sand and are therefore breccias.

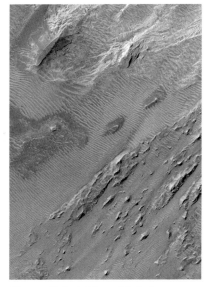

Figure 3.9 A Mars Global Surveyor image of a series of ripples and dunes in the Candor Chasma area of Mars for use with Question 3.8.

CHAPTER 4
THE MARTIAN LANDSCAPE

4.1 Planetary resurfacing

Compared with Mars, the Earth is a very dynamic world and has a relatively young surface. There is very little remaining that records the Earth's earliest past because the evidence has been obliterated mainly by the creation and destruction of crust through the action of plate tectonics (Box 4.2). There is no evidence of plate tectonics presently operating on Mars. However, this does not mean that Mars has a single uniformly old surface. In this chapter you'll learn how volcanism, which is part of the crucial driving force of plate tectonics on Earth, has acted in different ways on Mars to produce a varied geological surface to the planet.

The surface of the Earth is made up of a number of units with vastly different ages. When we say the surface is relatively young, it is the *average* age of the Earth's surface that is being referred to.

> Heat generated inside the planet and subsequently transferred to the surface and lost to space is one of the driving forces behind planetary resurfacing. Volcanism is the most common expression of that heat transfer and in this chapter you'll examine the processes that cause planetary volcanism.

- ■ Can you think of another process that will resurface a planet that you first met in Chapter 1?
- ❑ Impact cratering – the results of an impact with a planetary surface of objects travelling at many kilometres per second – is also one of the most significant factors affecting many planetary surfaces.

We'll also look at the role of impact cratering in modifying planetary surfaces and some of the other processes that modify the martian landscape later in this chapter.

4.1.1 The importance of heat

In the previous chapter, you examined the evidence for layering in both Earth and Mars and how the planets came to differentiate into separate layers through the effects of various heating processes. The processes by which the heating was, and continues to be, generated are outlined in Box 4.1.

In this section we'll consider how this heating might affect a planet once the core, mantle and crust have been formed. The heat is slowly lost through the surface, so cooling the planet. However, how does the heat generated within a planet get to the surface? Three main mechanisms operate within a planet such as Earth; these are conduction, convection and advection.

BOX 4.1 SOURCES OF HEAT

The most widely accepted theory for planet formation that you met in Section 3.2 requires that a planet needs to be heated before differentiation can occur and layering can begin to develop. There are several sources of heat that can operate during the evolution of a planetary body. The most important are:

- heat sources that develop during the early stages of planetary evolution, known as primordial heat sources (e.g. those associated with accretion and collision)
- heating processes that can operate long after the planet has formed (e.g. the heat generated by the decay of radioactive elements or the heat generated by the gravitational interaction of two planetary bodies).

Impact heating

During their formation and early existence, planets such as Mercury, Venus, Earth and Mars must have experienced heating due to impacts. However, the intensity of this bombardment would have waned over time as the growing planetary embryos collected the remaining debris. This final stage of bombardment was probably the result of capturing debris derived from earlier giant impacts during the accretion phase, together with an ever decreasing collection of planetesimals. Once the planets had become assembled, the later impacts would, on average, have been progressively smaller and less frequent. Material delivered during this later bombardment stage could, therefore, really only have delivered heat to the planet's newly formed surface crust.

Not all planetesimals were captured at this time; several thousand survive to this day as asteroids.

Tidal heating

One heat source known to be generated within some planetary bodies is **tidal heating**, which is created by the distortion of shape resulting from mutual gravitational attraction. Tidal effects are readily observed on the Earth's oceans where the attraction created by the Sun and Moon produce 'bulges' in the ocean water-masses, which are then dragged around the planet as the Earth rotates. This produces the ebb and flow of tides observed at coastal locations. In much the same way, the solid Earth is also distorted by these forces so producing 'tides', with a maximum vertical displacement of the rocky surface of up to 1 m. This deformation causes heating within the planet, though precisely where this heating is concentrated depends upon a planet's internal properties. In the Earth's case, it is thought to occur largely within the crust and mantle. Perhaps the most spectacular example of tidal heating is that seen on Io (see Figure 1.6), where tidal effects (created by the planet Jupiter) generate active volcanism.

Radiogenic heating

During the latter half of the 19th century, the eminent physicist William Thomson (Lord Kelvin, 1824–1907) attempted to determine the age of the Earth by considering that it had cooled slowly after its formation from a molten body. In effect, he was assuming the main sources of energy were from primordial heat and tidal heating. Taking many factors into account, including the mass of the Earth, the current rate of heat loss at its surface, and the melting points of various constituents, he concluded that our planet could not be much older than about 20–40 million years. We now know that the Earth is 4560 million years old from modern radiometric dating, a technique used for measuring the ages of rocks and minerals from the decay of certain radioactive elements within them. So why did Kelvin get the answer so wrong? The answer lies in the decay of certain unstable radioactive elements, the discovery of which was not made until some decades after Kelvin's initial calculations. It is now known that the energy released in this radioactive decay creates an important, independent source of heat within the Earth, which supplements that remaining from the primordial sources. This radiogenic heating is something that Kelvin could not possibly have known about when making his calculations. Radioactive decay has been a continuous source of heat within our planet since Earth's formation and, by analogy, also within other planets such as Mars.

Conduction

Conduction is perhaps the most familiar since it is the process of heat transfer experienced in the kitchen as heat is conducted from a stove to a pan, and ultimately to its handle. Different materials, such as rocks of various compositions, will conduct heat at different rates, and the efficiency of heat transfer in this manner is known as heat conductivity. This method of heat transfer is the most important in a planet's outermost layer (i.e. the crust).

Convection

The convection process relies on the fact that most materials expand when heated and, as a result, become less dense (i.e. more buoyant), allowing them to rise up through more dense surrounding material. Another example from the kitchen is the observation of swirling patterns in a pan of oil being heated prior to cooking. The patterns are the result of the movement of hot oil from the bottom of the pan upwards to the surface, accompanied by cooler oil sinking to the bottom.

It may surprise you to learn that silicate rocks can flow when the pressure and temperature become sufficiently high. The process itself is known as **solid-state convection** and, whilst rates may be no more than a few centimetres per year, it is the most efficient form of heat transfer. Near the surface of a planet the rocks will be too cold and rigid to permit convection, so conduction is the most significant process. The zone within the mantle of a planet in which both temperature and pressure are sufficient to permit flow is known as the **asthenosphere** (from the Greek *astheno* meaning weak), and this comprises much of the mantle thickness down to the core. Above the asthenosphere, the uppermost mantle and all of the overlying crust is solid and rigid and is given the name **lithosphere** (Figure 4.1). The division between the asthenosphere and lithosphere is based upon their physical properties and is not the same as the boundary between the mantle and the crust, which is primarily a compositional change.

> The lithosphere is an outer, rigid layer through which heat is transferred by conduction. The asthenosphere is an underlying, mechanically weaker layer that is capable of flow, and in which the principal process of heat transfer is convection.

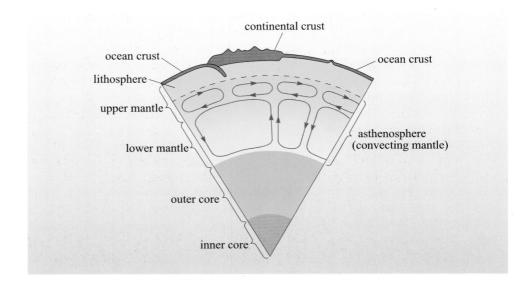

Figure 4.1 Section through the Earth showing the division of the mantle into the uppermost, rigid lithosphere, and the mobile, convecting asthenosphere (not to scale).

Advection

The final process of transferring heat is by physically moving molten rocky material, called magma, up through fractures in the lithosphere. This process is termed **advection** and operates whether the magma spreads out at the surface as a lava flow or, if it is injected, cools and crystallizes within the lithosphere itself. The effect is the same in both cases since heat is transferred by the molten rock from deeper levels where melting is taking place, to shallower levels where it solidifies, losing its heat by conduction into the overlying crust. Any planetary body that exhibits, or has exhibited, volcanic activity must have lost some of its internal heat in this manner.

QUESTION 4.1

Which *two* of the following are examples of heat transfer by convection?

A   Heat loss from the rear of a refrigerator to the surrounding air.

B   The upward movement of a parcel of warm atmosphere over a volcano on Mars.

C   The transmission of heat from the surface of a spacecraft to space.

D   The warming of a spacecraft en route to Mars by the Sun.

E   The transfer of heat though the walls of a house.

4.1.2 Plate tectonics on Mars?

Is resurfacing due to the loss of heat from the planet's interior still likely to be happening on Mars? On Earth, it is the convection in Earth's asthenosphere, which is a response to internally generated heat sources and remaining primordial heat (Box 4.1) that enables the transfer of this heat to the lithosphere, after which it is conducted and eventually lost into space. Resurfacing of Earth occurs largely as a result of volcanism (Section 4.2.1) or the movement of the lithosphere by plate tectonic processes (Box 4.2), for example where the collision of two lithospheric plates causes buckling and deformation.

The regions deformed in this manner form mountain belts such as the Alps or Himalayas. Over time, the mountains are eroded; the resulting small fragments are transported and deposited elsewhere, supplying the sediments that cover the lowlands and floors of ocean basins. If plate tectonics has occurred on Mars then similar styles of deformation at the plate boundaries might be expected. If heat loss from the interior has caused planetary resurfacing on Mars, then we need to examine the nature of the available heat sources on Mars and Earth. The efficiency of heat loss from Mars will differ from the Earth due to its smaller size (i.e. Mars has a smaller surface from which to lose heat) and because of differences in the relative thicknesses of the martian asthenosphere and lithosphere. Differences such as these can lead to substantial variation in the degree and nature of volcanic activity and the tectonic features that may be expected.

BOX 4.2 HEAT LOSS AND PLATE TECTONICS ON EARTH

The importance of the difference between the properties of the lithosphere and asthenosphere became apparent during the 1960s with the development of the theory of plate tectonics. This theory recognizes that the different parts of the Earth's lithosphere, comprising the non-convecting upper mantle together with the overlying oceanic or continental crust, can move relative to each other as a series of rigid plates (Figure 4.2). Current rates of movement are mostly between 50 mm and 100 mm per year, and the motion is thought to be a response to convection and heat loss in the mantle below the lithosphere. The movement of Earth's tectonic plates is enabled by the fact that new crustal material is added incrementally along **mid-ocean ridge** systems, and re-absorbed into the mantle at **subduction zones** that are associated with deep trenches at the edges of ocean basins. The main features of the outer layers of the Earth that give rise to plate tectonic motion are summarized in Figure 4.3.

■ What happens to ocean lithosphere over time as a consequence of its generation and destruction?

❑ A consequence of the generation and destruction of ocean lithosphere is that material is recycled within the upper mantle.

Plate recycling in this manner effectively adds hot material to the lithosphere at the mid-ocean ridges, and removes and re-absorbs cold, solidified material at subduction zones, and in doing so assists in the process of the surfaceward transfer of heat. A further consequence of tectonic recycling is that it can resurface large areas of a planet's lithospheric crust and, in the process, destroy the record of previous impact cratering or other surface-modifying processes.

Another topographic effect observed on Earth is the generation of broad surface uplifts, which produce bulges of hundreds of metres elevation and up to a 1000 km across in the overlying lithosphere. These are apparently unrelated to plate boundaries and are thought to be created by pipe-like zones of anomalously hot material rising up from deep within the mantle. The exact cause of these upwellings, or **mantle plumes** as they are more commonly called, is not yet known but they clearly represent another process by which heat may be transferred from deep within the Earth. Surface uplift occurs not only because there is a surfaceward movement of mantle material, but also because this rising mass of hotter mantle material expands due to the release of pressure during its upward journey. In addition, the release of pressure can result in the melting of the upwelling mantle, so such bulges are often associated with **hot spots** because the escape of this melted material produces volcanic activity. The best examples of present-day hot-spot volcanism related to mantle plumes are the volcanic islands of Hawaii and Iceland. Similar uplifted regions associated with volcanic constructions have been recognized on the surface of Mars in the form of the Tharsis 'bulge' (Section 4.2.4).

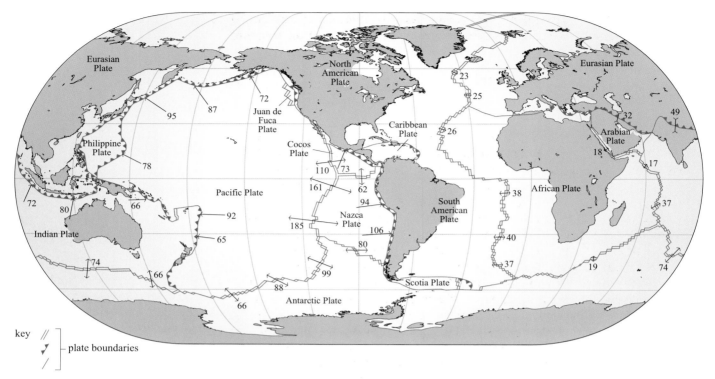

Figure 4.2 Map showing the global distribution of plates and plate boundaries on Earth. The black arrows and numbers give the direction and speed of relative motion between plates. Speeds of motion are given in mm yr^{-1}.

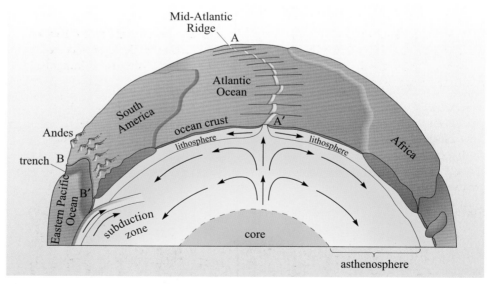

Figure 4.3 A vertically exaggerated model showing the main elements leading to plate tectonic movement on Earth. Since the lithosphere is rigid and comprises the crust and uppermost mantle, heat transfer is primarily by conduction, with a component of volcanically driven advection, i.e. the emplacement of magma into ocean lithosphere at mid-ocean ridges. The asthenosphere comprises the underlying convecting mantle. Lithospheric plates move apart at mid-ocean ridge systems, where material is continually added incrementally by volcanism (A–A′), and is destroyed at subduction zones where one plate is forced below another and reabsorbed into the asthenospheric mantle at depth (B–B′).

- Why will all planets gradually cool over time as a consequence of heat transfer and the loss of heat from their surfaces to space? (Hint: consider the nature of the internal heat sources.)

❑ Planets will cool because the sources of internal heat cannot be renewed (Box 4.1).

As cooling progresses, the lithosphere will thicken, and the top of the convecting asthenosphere will retreat inwards towards the centre of the planet. Mars is likely to have generated more modest amounts of internal heat than the Earth because of its smaller size so the cooling process and the thickening of the lithosphere will progress more rapidly on Mars than on Earth. Any associated volcanism or tectonic movement will also diminish. If the cooling process continues to the point where convection in the asthenosphere ceases altogether, all the outer layers of the planet will, by definition, become lithosphere. In such cases, the planet will become tectonically 'dead' and the associated resurfacing processes will effectively be terminated. However, even on those bodies where plate tectonic movement has ceased altogether, there may still remain sufficient internal heat to allow partial melting to occur. Consequently, this molten material may continue to generate surface volcanism even on 'tectonically dead' planets, provided the thickened lithosphere can be breached by the rising magma. Under these circumstances volcanic resurfacing, such as blanketing of earlier topography by lava flows or ash deposits, could continue even after surface tectonic processes have apparently ceased altogether. This appears to be the case on Mars as has been revealed by detailed studies of the martian surface by spacecraft.

4.1.3 The topography of Mars

Although Mars is a small planet – its radius is just a little over half that of the Earth – it has a dramatic topography, with a 30 km range in surface height compared to around 20 km for the Earth. The martian landscape has some of the most dramatic features in the Solar System.

There are superb examples of volcanoes on Mars, which, unlike Earth, do not occur in long chains or arcs. This suggests that the outermost layers of Mars (its crust and upper mantle) are not divided into mobile plates. Martian volcanoes, such as Olympus Mons that rises to a height of 24 km above the local surface, contribute to the 30 km range in surface elevation (Figure 4.4). The largest comparable structure on Earth, the volcano comprising the island of Hawaii,

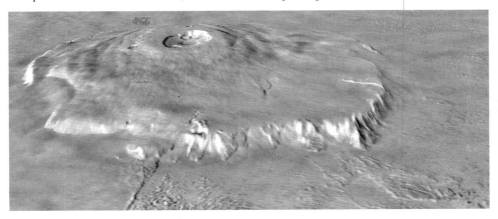

Figure 4.4 A computer-generated 3-dimensional view of Olympus Mons obtained by superimposing topography data from Mars Global Surveyor on an image obtained by one of the Viking orbiters.

is only one-hundredth the volume of Olympus Mons, yet its mass has created a substantial moat-like depression in the surrounding oceanic crust and upper mantle due to the shear mass of the volcano pushing the crust downward. Despite its size, no similar depression has been observed around Olympus Mons. This suggests that the height of martian volcanoes may indicate the presence of a very thick crust and upper mantle.

■ What is the thickness of the oceanic crust on Earth?

❏ From Figure 3.7, the Earth's oceanic crust is 6–11 km thick.

By comparison, the martian crust is believed to be 70–100 km thick.

You may find it helpful to have the *Topographic Map of Mars* to hand while studying this section.

Olympus Mons lies at the western edge of another significant feature of the martian landscape, the Tharsis Rise. This is a 10 km high, 4000 km wide bulge in the surface of Mars which is dominated by large volcanoes (Figure 4.5). Running from the eastern flanks of the Tharsis Rise, roughly along the equator, is Valles Marineris (Figure 4.6), a large canyon in the martian crust that is over 3000 km long (about one-fifth of the martian circumference), up to 600 km wide and 7 km deep. In the southern hemisphere of the planet is the Hellas Basin, which is an enormous impact crater that is 1800 km in diameter and more than 9 km deep (Figure 4.7).

Most of the detailed information we have about the topography of Mars, such as that shown in Figure 4.4, has come from an instrument on board the Mars Global Surveyor spacecraft called the Mars Orbiting Laser Altimeter (MOLA).

Figure 4.5 The Tharsis Rise and its associated volcanoes (mostly covered in bluish-white clouds) dominate the centre and left of this image obtained by the Mars Global Surveyor spacecraft in April 1999. To the lower right of the image is the Valles Marineris canyon.

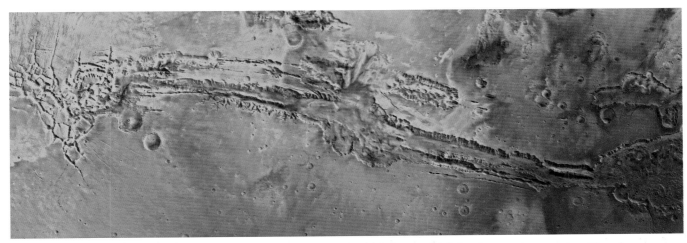

Figure 4.6 Viking orbiter image of the Valles Marineris canyon, which is over 3000 km long and on average 7 km deep.

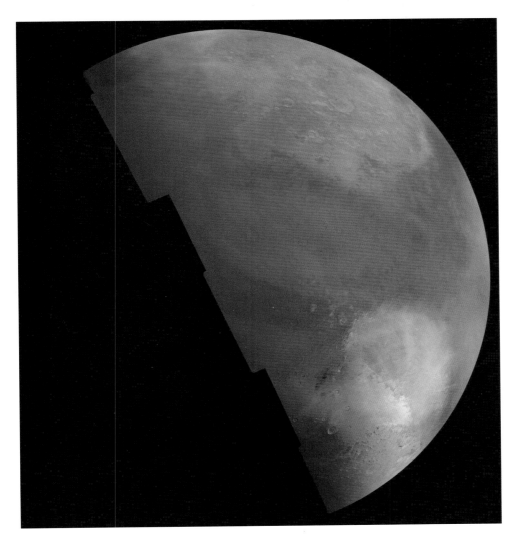

Figure 4.7 A Viking orbiter image of the Hellas Planitia region of Mars; north is towards the top of the image. The scene shows the 1800 km diameter Hellas Basin (covered by clouds), an ancient impact basin (and the largest basin on Mars) formed when a large projectile (asteroid, comet or meteor) hit the surface.

This has enabled detailed studies of the martian surface that suggest there are no obvious plate tectonic boundaries in the lithosphere, although there appears to be two distinct types of crust, which may be broadly analogous to the continental and oceanic crusts of Earth (Box 4.2). The southern hemisphere and part of the northern hemisphere form a huge ancient region of heavily cratered highlands (Figures 4.8 and 4.9). This elevated, older crust could be compared with Earth's continental areas. By contrast, the northern hemisphere is on average 6 km lower, much less heavily cratered, and is apparently covered with lava flows and sedimentary material.

■ What might a lack of craters on one hemisphere relative to the other indicate about the extent of resurfacing?

❏ It could indicate that the hemisphere with fewer craters has seen more resurfacing.

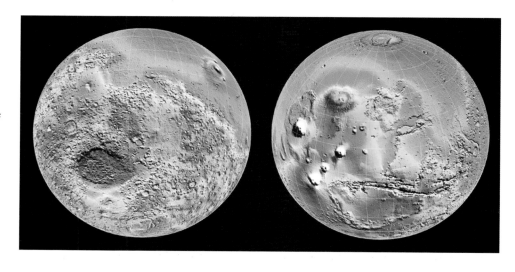

Figure 4.8 The differing hemispheres of Mars. False-colour topographic maps of the southern (left image) and northern (right image) hemispheres of Mars. The colours represent the relative heights of the landscape: dark-blue represents the lowest terrains and reds the highest; the scale for the colours is shown in the global map in Figure 4.9.

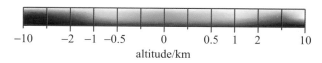

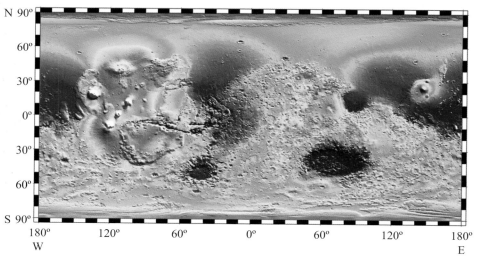

Figure 4.9 A map of Mars' global topography constructed from data obtained by the Mars Orbiting Laser Altimeter on board the Mars Global Surveyor spacecraft. The colours represent the relative heights of the landscape: dark-blue represents the lowest terrains and reds the highest. The scale for the colours is shown above the map and gives the relative altitudes in kilometres.

Some areas of Mars also have magnetic stripes that are similar to those produced in oceanic crust on Earth as a result of plate tectonics (Figure 4.10). On Earth, magnetic stripes such as these are preserved in oceanic crust as a result of changes in the Earth's magnetic field over time, each stripe representing a major change in the Earth's magnetic field. As oceanic crust spreads away from mid-ocean ridges, the newly formed crust preserves the state of the Earth's magnetic field at that time, giving rise to a series or parallel magnetic stripes. The observation of similar features on Mars has led to speculation that there may have once been some form of plate tectonics on Mars, and that these regions are thus analogous to oceanic crust on Earth. However, the martian crust and mantle have thickened significantly since they were formed due to the cooling of the planet, to the extent that any transfer of heat within the mantle must occur at depth. Consequently, the only surface expression of surfaceward transfer of internal heat is a global distribution of volcanoes, and the presence of elevated surface regions such as the Tharsis Rise (Figure 4.5). The elevated Tharsis region, upon which large volcanoes including Olympus Mons are located, is believed to be the expression of a particularly persistent source of heat within the martian mantle in much the same way as localized sources of heat within the Earth's mantle, called mantle plumes, are associated with the construction of major volcanic structures (e.g. Hawaii). Nevertheless, volcanic activity on Mars does appear to have waned.

From cratering studies, Mars' highland crust appears to have originated about 4.5 billion years ago. However, this ancient terrain has been substantially modified by later volcanic activity. The volcanic lowlands in the northern hemisphere, and the equatorial Tharsis region probably began to develop no earlier than 3 billion years ago, and some lava flows here may be no older than 10 million years. Martian volcanic activity cannot yet be considered as entirely extinct.

There are no absolute dates for the martian surface. Ages from cratering studies are calibrated with equivalent dates from the Moon where radiometric dates are obtained by laboratory studies of lunar rocks obtained from the Apollo missions.

QUESTION 4.2

In four or five sentences, summarize the main points for and against some form of plate tectonics on Mars at the present time or at some point in the planet's past.

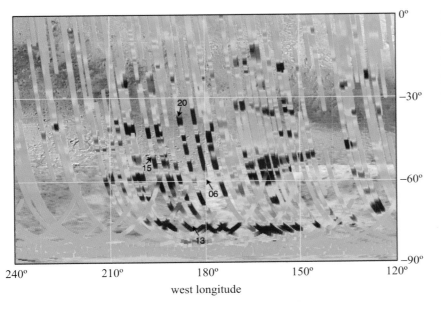

Figure 4.10 This false-colour image obtained by Mars Global Surveyor is a map of martian magnetic fields in the southern highlands where magnetic stripes, possibly resulting from crustal movement, are most prominent. The bands are oriented approximately east–west and are about 150 km wide and 900 km long. The false blue and red colours represent magnetic fields in the martian crust that point in opposite directions. The magnetic fields appear to be organized in bands, with adjacent bands pointing in opposite directions, giving these stripes a striking similarity to patterns seen in the Earth's crust at the mid-oceanic ridges.

4.2 Volcanism on Mars

4.2.1 What is volcanism?

Volcanoes are the surface expression of the melting processes that occur deep within a planet. The volcanic 'plumbing' systems beneath the volcano link these deep processes to the surface and any atmosphere that might be present.

> Volcanism can be defined as the processes associated with the transfer of molten material (magma), associated volatile material and any suspended load of crystallized material, from the interior of a planet to its surface.

On Earth, volcanic processes are most obviously demonstrated by the presence of volcanoes that erupt lavas and produce ash clouds and toxic gases. There are many types of volcanoes on Earth – different styles of volcanic activity are largely controlled by the chemical and physical properties of the magma. As a result, volcanic activity ranges from relatively quiescent pouring out of lava, such as might be seen on Hawaii (Figure 4.11), to massively explosive eruptions such as the 1980 eruption of Mount St Helens (Figure 4.12). We'll look at the styles of volcanism in more detail in Section 4.2.2. Yet, for any type of volcanic eruption there are two prerequisites: a process that causes melting and material that can be melted.

Figure 4.11 A quiescent lava eruption, an example of effusive volcanism. Lobes of lava advance quietly across an earlier lava flow. This basalt eruption took place during 1999 on the lower flanks of the Kilauea volcano in Hawaii, and was accompanied by only minor ash and gas release.

Figure 4.12 A volcanic explosion. The eruption of Mount St Helens, Washington State, USA. At 8.32 am on 18 May 1980. Part of its flank slid away in a gigantic rockslide and debris avalanche, releasing pressure within the volcano. This triggered a major explosion and the generation of a huge eruption column.

4.2.2 Styles of volcanism

The nature and styles of volcanism vary greatly, and range from relatively quiescent lava eruptions with relatively little or no volcanic ash, to mighty explosions that create immense clouds of hazardous ash and gas that can profoundly affect the environment. Volcanologists recognize these two contrasting styles of activity as **effusive volcanism** and **explosive volcanism** (Figures 4.11 and 4.12).

> Effusive volcanism is typically a quiet affair since it is characterized by lava emanating from a vent or fissure and then spreading out over the landscape. By contrast, explosive volcanism typically produces fragmented debris or **pyroclastic materials** (from the Greek *pyro*, meaning 'fire-broken'), which largely comprise ash and other ejected fragments.

Large, extremely violent and voluminous pyroclastic eruptions have been associated with some of the major natural catastrophes in Earth's history. For instance, the eruption of Vesuvius in AD 73 buried the Roman city of Pompeii, whilst the effects of the 1620 BC explosion of the volcanic island of Santorini in the Aegean are thought to have destroyed the flourishing Minoan civilization. More recently, the eruption of Krakatau (Indonesia, 1883) as well as more modern examples such as eruptions of Mount Pinatubo (Philippines, 1991) and Mount St Helens (USA, 1980) have also had wide-ranging effects on communities and measurable effects upon global climate.

The way in which magma is generated and makes its way to the surface, together with the style of the resulting volcanism, are controlled by a range of physical factors. These factors will be discussed in the following sections.

Effusive volcanism

When magma reaches the surface of a planet and flows across it to form a channel or a sheet it is called a **lava flow**. Erupted lavas are typically liquids, but they can also contain crystals, which form during cooling as the magma rises, and gases dissolved in the magma, which begin to form bubbles as the pressure is reduced once the magma nears the surface. Differences in magma composition and the amount of gas it contains control how explosive the resulting volcanism will be since these factors influence an important property of the magma: its viscosity.

> Viscosity is a measure of the resistance of a fluid to flowing. Water has a lower viscosity than engine oil, which in turn has a lower viscosity than treacle. A lower viscosity gives a lava the ability to flow more easily.

For example, when a liquid lying on an inclined surface begins to flow downhill, two forces control the rate at which the liquid can spread: gravity and viscous resistance. In addition to the viscous resistance of the liquid, molten lavas also form a chilled crust on their surfaces (Figure 4.13) so that before a lava can flow the thin crust must give way (i.e. it must 'yield' and begin to deform), a physical property known as the yield strength of the lava. Before a lava can spread, the yield strength must be exceeded by forces which are acting to make the lava flow (e.g. gravity).

Whether or not a lava will flow readily at the surface of a planet depends upon factors such as the strength of gravity, and surface and eruption temperatures, which will control the cooling rate of the lava and its chemical composition. Magmas with a basaltic composition produce particularly fluid lavas at temperatures above 1050 °C.

Figure 4.13 Low-viscosity basalt lava oozing out and advancing as a series of 'toes', which build and coalesce to form laterally extensive lava-flow sheets.

Lavas, in common with many other fluids such as motor oil, become more runny (i.e. their viscosity decreases) with increasing temperature. Conversely, they will only remain runny if they continue to remain at temperatures well above their melting point (Figure 4.13). In other words, a magma erupted well above its melting point is more likely to produce a lava flow than one erupted at only a few degrees above its melting point since the latter will solidify before it can travel far. In the first instance, a thin, laterally extensive lava flow would be formed because the magma can spread outwards from the vent, whilst a taller mound or dome-like construction might be created around the vent in the second instance. The surface temperature will also play an important role since, if the surface conditions of a planet are significantly cooler than the temperature of erupting lava, it will be less likely to flow far due to rapid chilling.

Eruption rate is another factor known to affect the pattern and structure of lava accumulations. Where magma rises and oozes slowly onto the surface, it rapidly chills and solidifies. By comparison, at high rates of eruption the supply of magma is so rapid that chilling cannot occur efficiently, and the lavas can flow greater distances and cover a wider area.

Effusive eruptions give rise to the largest outpourings of lava on Earth, known as flood basalts. These can produce volumes of up to 2 million cubic kilometres over a period of 1–5 million years. One example is the Colombia River basalt province of North America which was erupted 14–15 million years ago (Figure 4.14).

Explosive volcanism

Explosive volcanism is largely the result of two main properties of a magma: its viscosity and the amount of gas it contains. Basaltic lavas generally have low viscosity and relatively low gas contents, both factors working together to produce relatively quiescent, effusive basaltic eruptions. Where gas contents are higher in basaltic magmas, **fire fountains** may occur casting lava upwards for several tens

Figure 4.14 Layers of stacked lava flows in the Columbia River province, covering parts of Washington, Oregon and Idaho. Much of this province was erupted between 14 million and 15 million years ago, and contains some of the largest lava flows yet identified on Earth. The flows shown represent just a small thickness of the voluminous and rapidly erupted flood basalt.

Figure 4.15 A fire fountain during the eruption of Pu'u'O'o, Hawaii.

to hundreds of metres. Spectacular examples of fire fountains occur from time to time on Hawaii (Figure 4.15). Nevertheless, basalt fire fountains are small affairs compared with major explosive eruptions. In these instances, the expanding magmatic gas not only fragments the erupting lava but also drives it upwards creating **eruption columns** reaching heights of tens of kilometres above the vent on Earth, and even higher on other planetary bodies with weaker gravity (Figure 4.16).

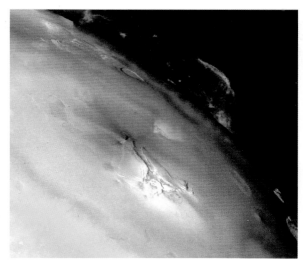

Figure 4.16 The eruption of Pele volcano on Io, a satellite of Jupiter. The volcano is around 35 km across and, because gravity is weak on Io, eruption columns can reach heights of 70–280 km above the surface. In this image the central vent of the volcano is the pale-coloured area in the centre of the image. The plume of material dispersed by the eruption column is 1000 km in diameter and shows up against the blackness of space.

So, what causes the differences between these effusive and explosive styles of eruption? To understand this, it is necessary to consider what happens to magma as it approaches the surface. A useful analogy is to consider the gentle or rapid escape of dissolved gas when a new bottle of fizzy drink is opened, particularly if the bottle has been shaken a bit first to simulate the movement of magma.

■ What will happen if the bottle is opened (a) slowly and (b) quickly?

❑ (a) If the bottle is opened slowly, the gas has time to escape and the drink can be poured without spillage; (b) if opened quickly, the pressure release causes bubbles to form very rapidly producing a froth which then explodes from the bottle spout.

In the case of magma, its ascent to the surface of a planet is driven partly by buoyancy (partial melting typically produces liquids of different composition and lower density than the source) and partly by pressure (liquids will flow towards the surface away from the higher pressures at depth). The magma's buoyancy is further enhanced because, as it rises, it will be hotter than the surrounding rocks. As with our fizzy drink, as the magma ascends the pressure confining it is reduced allowing any gases dissolved in the magma to expand and form bubbles (Figure 4.17). This bubble formation serves to make the magma even less dense and therefore even more buoyant, thus further accelerating its ascent. Bubbles continue to expand as they rise in response to further decreases in pressure until they reach the surface where the gases escape into the atmosphere. If the magma is not viscous, then bubble escape may take place in an uninhibited, gentler fashion, producing lavas with preserved bubbles, or in those cases where the gas content is higher, causing fire fountaining.

However, if the lava is relatively viscous this expansion and release of gas cannot take place as easily. As a result, the magma reaches the surface containing highly pressurized bubbles that have been unable to expand fully and escape during magma ascent. Once beyond the confines of the lava conduit, the gases within the erupting lava expand rapidly creating an explosion.

Violent volcanic eruptions are much more common in volcanoes with more viscous and gas-rich magmas. The explosions they create literally blow apart the ascending magma, creating semi-molten clots of lava known as **volcanic bombs**, together with pumice, **spatter cones** and ash. These are all pyroclastic materials, since they are produced due to fragmentation of the magma created by gas expansion.

Spatter cones are small (5–20 m high) volcanic cones built from material blown out of volcanoes as clots of relatively fluid basaltic lava.

QUESTION 4.3

From the following list, which *one* statement is a correct reason why the Earth is more volcanically active than Mars?

A Tidal heating of the Earth by the Moon.

B The advection of magma into the Earth's continental land masses.

C The lack of a strong magnetic field on Mars.

D Mars has cooled more quickly than the Earth.

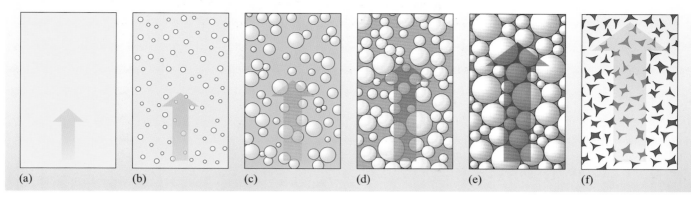

Figure 4.17 Effects of the loss of gas from a magma in a volcanic conduit: (a–b) as the magma begins to ascend in the conduit bubbles begin to form, so enhancing the rise of the magma; (c–d) with continued ascent in the conduit additional reduction in pressure allows a greater amount of degassing, and so leads to the formation of more bubbles which begin to coalesce into larger bubbles and further increase the buoyancy; (e) as the rising magma accelerates in the conduit, large amounts of degassing, or else rapid degassing within more viscous magmas, can result in fragmentation and the production of pyroclastic materials (f). Stages (a), (b) and (c) are most typical of the degassing characteristics of less viscous basaltic lavas, or else gas-poor magmas, whereas stages including (d), (e) and (f) are more typical of the degassing of gas-rich, or highly viscous magmas.

4.2.3 Volcanic landforms on Mars

We've already seen that the tremendous size of martian volcanoes is one of the consequences of the thickness of the martian lithosphere and its ability to support the immense mass of volcanoes such as Olympus Mons.

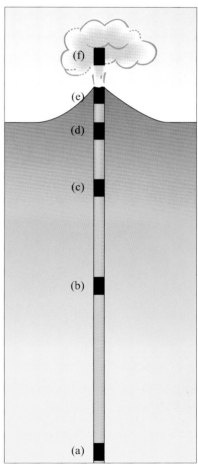

- ■ Plate movements do not currently take place on Mars. Can you suggest how this might also contribute to the fact that martian volcanoes can be much larger than Earth's largest volcanoes?

- ❑ The Hawaiian volcanoes are moved away from their underlying heat sources by plate tectonics, thus limiting the maximum size they can achieve. By contrast, a volcano above a heat source on Mars remains fixed, enabling it to grow indefinitely in size.

A number of volcanic landforms have been identified in high-resolution images of the martian surface that have volcanic counterparts on Earth.

Montes

Montes (singular Mons) are sometimes simply referred to as **shield volcanoes** (Figure 4.18). They have a morphology similar to the Hawaiian shield volcanoes on Earth, but are far greater in size. They are characterized by gentle slopes and broad **calderas** at their summits. Olympus Mons (Figure 4.19), the largest of this type of feature, has a base diameter of over 600 km and a summit caldera complex more than 60 km in diameter and 3 km deep. Its summit towers some 24 km above the surrounding plains.

Calderas are large, volcanic depressions formed by the collapse of the rocks around the mouth of a volcano.

Arsia Mons is located at 10° S, 240° E on the *Topographic Map of Mars*.

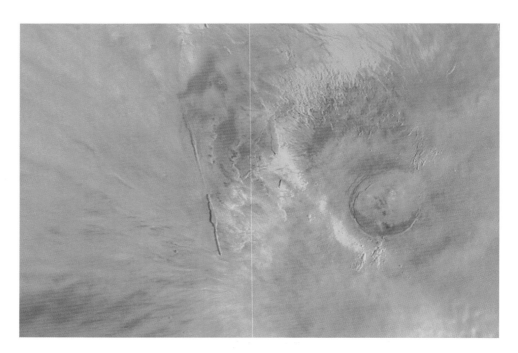

Figure 4.18 A Mars Global Surveyor image of Arsia Mons, one of the largest volcanoes on Mars. This shield volcano is part of a group of volcanoes known as the Tharsis Montes, the others are Pavonis Mons and Ascraeus Mons. The summit of Arsia Mons is more than 9 kilometres higher than the surrounding plains and its caldera is approximately 110 km across.

Olympus Mons is located at 19° N, 225° E on the *Topographic Map of Mars*.

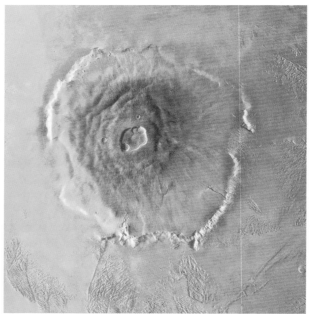

Figure 4.19 A spectacular view of Olympus Mons, the youngest volcano in the planet's volcanic Tharsis region. It is thought that it was last active over 100 million years ago. This volcano is 24 km high – three times taller than Mount Everest – and as wide as the entire Hawaiian Island chain. The three craters in the summit region are up to 3 km deep, and are thought to be caldera collapses formed by the drainage of lava beneath during eruptions from the flanks of the volcano. Despite its large size, the angle of slope is only a few degrees, similar to those of Mauna Loa, Hawaii. Consequently, it has long been inferred that this volcano was built by the repeated eruption of low-viscosity basaltic lava flows. The martian crust is able to support such a large volcanic edifice because, being a smaller planet than Earth, it has lost much of its heat which has enabled its lithosphere to thicken and so become much stronger.

Paterae

Paterae (singular Patera) are extremely low, flat shield volcanoes (Figure 4.20) with calderas that are often irregular and subtle when compared to the other types of large shield volcanoes on Mars. The summits of the paterae typically rise, at best, only a few kilometres above the surrounding plains. In spite of their low relief, they are often very large structures. In fact, Alba Patera in the Tharsis region (Figure 4.21) has a base diameter of 1200 km, significantly greater than that of Olympus Mons.

Tholi

Tholi (singular tholus) are smaller than the montes, though still impressive by Earth standards. They are dome-shaped with large summit calderas and relatively steep flanks (Figure 4.22).

Figure 4.20 A Mars Global Surveyor image of the Apollinaris Patera volcano. This ancient volcano (located at 9° S, 174° E) near the equator and is thought to be as much as 5 km high. The caldera, the semi-circular crater at the volcano summit, is about 80 km across. The channel-like patterns emanating from the summit caldera are thought to have been produced by erosion and deposition of pyroclastic material cascading down the flank of the volcano. The pale-blue area is a patch of bright clouds hanging over the summit.

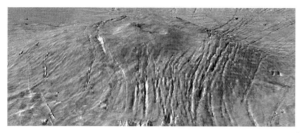
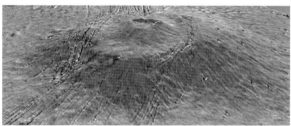

Figure 4.21 Two computer-generated 3-dimensional views of the volcano Alba Patera, produced by superimposing image data from the Viking orbiter on laser altimeter topography data from Mars Global Surveyor. The vertical height of the volcano has been exaggerated. Alba Patera is located at 40° N, 250° E.

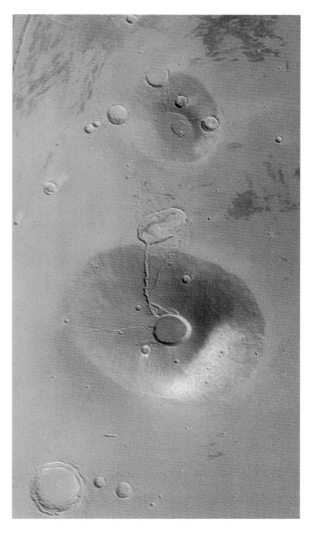

Figure 4.22 A Mars Global Surveyor image of the martian volcanoes Ceraunius Tholus (lower) and Uranius Tholus (upper). The presence of impact craters on these volcanoes, particularly on Uranius Tholus, indicates that they are quite ancient and are not active today. The crater at the summit of Ceraunius Tholus is about 25 km across and located at 24° N, 263° E.

Small shield volcanoes

The high-resolution images obtained by the Mars Global Surveyor spacecraft have revealed numerous small shield volcanoes (Figure 4.23) with diameters less than 10 km. Many of these show strong similarities to small shield volcanoes on Earth.

Cones and buttes

High-resolution images have also revealed dense fields of possible small cinder cones and buttes (Figure 4.24), many less than 1 km in diameter, similar to those found in volcanic areas of the Earth. Buttes are high, isolated, steeply sided hills and are common features of the landscape of the western United States, from where the term derives.

Lava flows

High-resolution images show innumerable lava flows on the slopes of martian volcanoes (Figure 4.25). Some of the flows are quite large, with estimated volumes of perhaps 100 cubic kilometres. In addition, vast portions of Mars' northern plains are blanketed with flood basalts. Rather than issuing from the great volcanoes, many of the flood basalts erupted from fissures and numerous small shield volcanoes. There is also some evidence for pyroclastic deposits on Mars.

4.2.4 Volcanic provinces on Mars

You saw in Section 4.1.2 that whilst there is some evidence for plate movements on Mars in its early history (Figure 4.10), it was probably not a process that lasted for very long. Mars is effectively a single-plate planet. However, there are two distinct types of terrain on Mars: the southern highlands, typified by rough, heavily cratered terrain, and the northern plains, which display a much younger face, with extensive evidence of widespread resurfacing (Figure 4.8). An escarpment 1–3 km in height bounds these regions. The volcanic features on Mars are not randomly distributed. The vast majority are located in the northern plains and, within this region, volcanoes are clustered into distinct provinces.

The Tharsis Rise is the greatest of the northern plains volcanic provinces, covering about one-quarter of the planet's surface area. The 'bulge' gently rises to a height of 10 km above the Mars datum, making it the highest area on the planet. Atop the 'bulge' are Mars' greatest volcanoes. The other major volcanic province of the northern plains is Elysium Planitia. Major volcanoes of Elysium include Elysium Mons, Albor Tholus, Hecates Tholus, and Appolonius Patera. The region also includes a number of smaller volcanic features. The volcanoes of the Elysium region sit atop a bulge similar to, but smaller than Tharsis, rising 2–3 km above the Mars datum.

The northern plains also exhibit several minor volcanic provinces (Table 4.1). Volcanoes of the southern highlands are few and scattered. Seven paterae have been identified: Nili, Meroe, Tyrrhena, Hadriaca, Amphtrites, Peneus, and Tempe.

Figure 4.23 A Mars Global Surveyor image of a small shield volcano (about 5 km across with a 1.2 km elongate depression at its summit) in the Tempe Terra region of Mars.

The Mars datum is the equivalent to 'sea-level' on Earth, i.e. the zero point of elevation. Since Mars has no ocean, it is defined as the point at which atmospheric pressure is 6.1 millibars. However, since martian atmospheric pressure can vary, the datum is also defined using more sophisticated gravitational parameters.

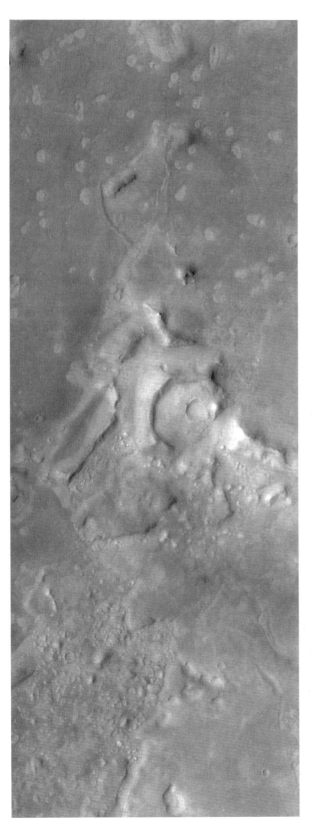

Figure 4.24 A Mars Odyssey image of the Acidalia Planitia region of the northern lowlands of Mars (centered on 50° N, 330° E). Much of this region is thought to be composed of relatively young volcanic flows. The lighter spots in the top portion of the image are believed to be small volcanic features, such as small cones. The appearance of the darker material abutting or flowing against the higher knobs and ridges is common in volcanic flow fields.

Figure 4.25 Lava flows on the southern flank of Olympus Mons.

Table 4.1 Minor volcanic provinces in the northern plains of Mars.

Volcanic province	Features
Arcadia–Amazonis	small domes and cones
Acidalia–Chryse	small domes and cones
Acidalia Planitia	small cones and domes, many with craters
Chryse Planitia	small shields and buttes
Isidis Planitia	long chains of coalescent crater cones that may be analogous to the crater and cone chains that form along fissures in rift zones on Mauna Loa in Hawaii
Utopia Planitia	numerous flows, cratered cones
Phlegra Montes	small cones and flows
Borealis region	few small cratered cones

There is compelling evidence that the vast majority of martian volcanism is basaltic in nature. Nearly all of the volcanic landforms on Mars you saw in Section 4.2.3 have basaltic analogues on Earth. The shield volcanoes are similar in morphology (if not in size) to the basaltic shield volcanoes of Earth. The presence of small cones and buttes are also consistent with similar features found in basaltic volcanic terrains on Earth. Because of the absence of plate tectonics on Mars there are none of the linear chains of volcanoes found around plate boundaries on Earth. Instead, Mars appears to exhibit hot-spot volcanism resulting in the development of shield volcanoes similar to some of the basaltic landforms found on Earth. Volcanic regions on Mars exhibit immense numbers of lava flows that exhibit structures and evidence of low viscosity that are consistent with basaltic flows. Finally, remote sensing of the martian surface by the Mars Odyssey spacecraft has enabled detailed mapping of the mineralogy of areas of the martian surface. Figure 4.26 shows one such image of an area known as Ganges Chasma in the Valles Marineris – the large equatorial canyon on Mars. This false-colour image (recorded in the infrared since different rocks and minerals emit infrared light to differing extents) shows compositional variations in the rocks exposed in the wall and floor of Ganges Chasma (blue and purple in Figure 4.26) and in the dust and sand on the rim of the canyon (blue and orange). Blue and purple areas are basaltic in composition; the purple areas are particularly rich in the mineral olivine, a common mineral constituent of basalts.

4.3 Impact craters

4.3.1 Impact cratering in the Solar System

In examining volcanism on Mars in the previous section, you've been looking at an aspect of the martian landscape that is relatively familiar to us here on Earth. The driving forces behind martian volcanism are primarily processes internal to the planet, i.e. heat and its transfer to the martian surface. However, a glance at many images of Mars reveals one feature of the martian landscape that, while not uncommon on Earth, is certainly prevalent on Mars: impact craters (Figure 4.27).

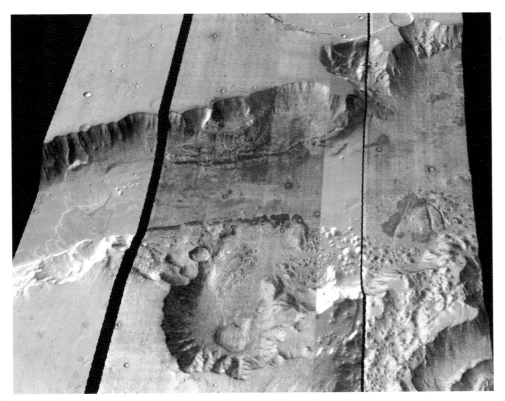

Figure 4.26 False-colour infrared image of the Ganges Chasma (8° S, 312° E) area of Valles Marineris on Mars obtained by the Mars Odyssey spacecraft. The floor of Ganges is covered by rocks and sand composed of basaltic lavas shown in blue. A layer that is rich in the mineral olivine can be seen as a band of purple in the walls on both sides of the canyon, and is exposed as an eroded layer surrounding a knob on the floor. Olivine is easily destroyed by liquid water, so its presence in these ancient rocks suggests that this region of Mars has been dry for a very long time. The image is approximately 150 km across.

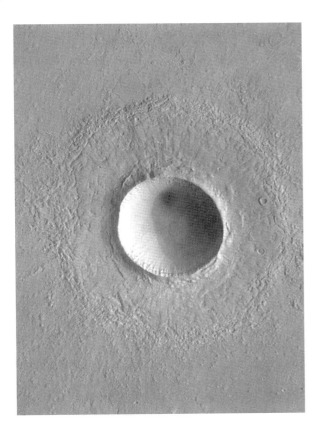

Figure 4.27 An as yet un-named impact crater, approximately 1.5 km across, in the Elysium Planitia region of Mars.

It may surprise you to learn that there are probably about 1000 objects over 1 km in diameter and possibly as many as a million over 50 m in diameter whose orbits cross, or closely approach that of the Earth. These objects are a mixture of asteroids and comets. In our voyage around the Sun, we travel through a 'shooting gallery', and the same is true for all the planets and satellites in the Solar System. Two to three of those objects, about 1 km in diameter, hit our planet every million years. Although on the scale of a human lifetime impacts are rare, from a geological perspective they are a common occurrence. This is the reason why **impact cratering** is by far the most widespread process shaping the surfaces of solid bodies in the Solar System.

Impacts are intimately associated with the formation and evolution of planets. Although some bodies in the Solar System were formed directly from the dust and gas in the primordial solar nebula (the Sun and giant planets fall into this category), almost everything else was constructed by impact and accretion of solid objects in the early Solar System. This impact 'cascade' began with sub-millimetre objects that accreted to form bodies that ranged from about one metre to tens of metres in size. These then accreted to form bodies about one kilometre to tens of kilometres in size, which themselves accreted to form planetary embryos. Current models for the formation of the terrestrial planets indicate that the final stages of planetary accretion are characterized by collisions between tens and hundreds of Mercury- or Mars-sized planetary embryos (see Section 3.3).

> It is entirely possible that life on Earth arose several times during the first 700 million years of our planet's history. However, continual large impacts would have caused global sterilization of the planet, frustrating the proliferation of life.

Very large impacts appear to be a fundamental part of planet formation. One such impact very early in the history of the Solar System, between a newly formed Earth and a Mars-sized body, is the preferred model for the formation of the Moon. When scientists look at the rate of large impacts over time since the origin of the Solar System, they find (thankfully) that there was a peak about 4.5 billion years ago, falling dramatically during the first few hundred million years of Solar System history to practically nothing today. However, around 3.9 billion years ago there was an apparent 'burst' of impact activity. This last 'burst' of large impacts is known as the late heavy bombardment and it was responsible for reshaping the lunar surface, and possibly frustrating the proliferation of life on Earth. Impacts with asteroids several hundreds of kilometres in diameter would have vaporized any early oceans on the Earth. Impact cratering remains a significant process on Earth, reshaping the surface and probably causing occasional mass extinction events, but there are very few obvious craters on the Earth's surface that are visible from space. This is because the surface of the Earth is continually being re-shaped by erosion, volcanism, and crustal deformation and destruction. These processes may be non-existent, less effective or, in some cases, have long since ceased on other planets, where craters are the dominant landform.

- ■ If impact cratering is a ubiquitous process, why does the Earth show so little evidence of it?

- ❏ On Earth, several geological processes – erosion, volcanism, crustal deformation and destruction – have been at work, wiping out all traces of the early heavy bombardment and evidence of many other large impacts since then (Figure 4.28).

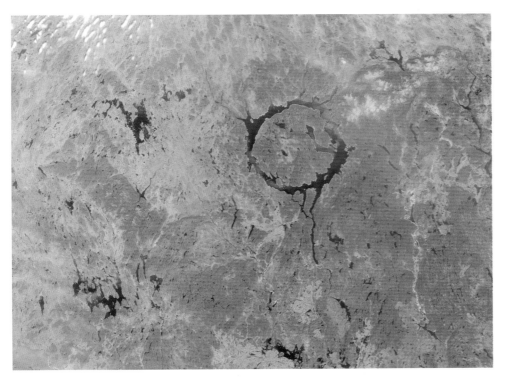

Figure 4.28 The large circular lake in this image represents the remnants of one of the largest impact craters still preserved on the surface of the Earth. Lake Manicouagan in northern Quebec, Canada, surrounds the central uplift of the impact structure, which is about 70 km in diameter and is composed of impact-brecciated rock.

Impacts that produce craters on terrestrial planets occur at speeds of several kilometres per second (10 km s^{-1} is equivalent to 36 000 km/h or about 22 400 mph). Relative encounter speeds vary throughout the Solar System – asteroidal material at Mars has an average impact speed of 10–11 km s^{-1}. The average encounter speed at the top of Earth's atmosphere is about 17 km s^{-1} for asteroidal material, and up to about 70 km s^{-1} for comets.

Because of these high speeds, impact cratering is the most rapid geological process known. Asteroids traverse the Earth's atmosphere in seconds and, following impact with the planet's surface, can form craters that are hundreds of kilometres in diameter in minutes.

4.3.2 The impact process

When objects less than 10–20 m in size pass through the Earth's atmosphere they lose most of the speed that they had in space because of air resistance, and lose much of their mass by ablation (i.e. loss of surface layers due to melting and vaporization), and hit the surface at a terminal **free-fall speed** of a few hundred metres per second. If the object is initially smaller than about 10–20 m then the atmosphere is sufficient to decelerate it significantly, since smaller objects have a higher surface area (and thus present a relatively larger area to the atmosphere) compared to their mass. Impact of these meteorite-sized objects may excavate a small pit, usually between 10 mm and 10 m across (see Figure 4.29) but in exceptional cases reaching a few tens of metres across.

Figure 4.29 A suburban setting for one of the smallest terrestrial impact features. This pit was formed by a meteorite that fell at Barwell, Leicestershire, 24 December 1965.

QUESTION 4.4

Study the photograph in Figure 4.29 of the pit produced by a relatively recent British impact. What can you say about the likely speed of the impactor, and the angle of its approach? How do these account for the appearance of the impact site? *Hint*: think about the size of the object, and the likely effects of the atmosphere.

In the case of larger objects, a few tens of metres or more in diameter, the situation is very different. These projectiles retain a large portion of their cosmic speed despite passage through the Earth's atmosphere, and hit the surface at speeds of several kilometres per second or, in some cases, several tens of kilometres per second. Of course, on planets or moons without an atmosphere, all objects whatever their size, strike the surface at such high speeds.

After a high speed impact, shock waves radiate out from the point of impact, moving huge volumes of target rocks, and creating true impact craters. The pressures involved in impacts are much higher than those involved in normal geological processes, typically over 1000 times higher.

A complex sequence of events and processes occurs during the formation and immediate modification of a new crater. These may be grouped into three broad stages of impact cratering:

- contact and compression
- excavation
- modification.

Figure 4.30 illustrates schematically the stages of the impact process during the formation of a simple crater.

Contact and compression stage

This first stage of impact cratering begins the moment the projectile makes contact with the ground surface (Figure 4.30a). As the projectile travels into the target, it compresses the target material and accelerates it to high velocities. Simultaneously, the projectile itself decelerates. Shock waves originate at the point where the projectile touches the target surface. Shock pressures during this initial stage reach extremely high values – a rock would have to be buried to more than 2000 km to experience the same kinds of pressure within the Earth. Both target and impactor are vaporized or melted when the pressure is released. The energy that the projectile has due to its motion, called its **kinetic energy**, is largely transferred to the target and heats, deforms and accelerates the target rocks. The release of pressure within the projectile effectively ends the contact and compression stage (Figure 4.30b). After this point, the projectile itself plays no further part in the formation of the final crater. This stage lasts for less than one second in all but the largest impacts.

Excavation stage

As the contact and compression stage ends, a roughly hemispherical shock wave surrounds the projectile and propagates into the target – the centre of this hemisphere actually lies well below the original ground surface, since the projectile may have penetrated up to twice its own diameter into the target. This initial shock wave, and other waves that are reflected from the original ground surface, weaken, fracture, and shatter the target rock. These waves also move material, excavating material from around the centre of the developing structure.

The movement of material upwards and outwards at upper levels, and downwards and outwards at lower levels, eventually opens out the crater, producing a bowl-shaped **transient cavity** that is many times larger than the diameter of the projectile (Figure 4.30c and d). Excavated material is ejected over the surrounding terrain. A point is reached when the various shock waves associated with the impact can no longer displace rock. The transient cavity reaches its maximum extent at this point. This marks the end of the excavation stage and the beginning of the modification stage.

The excavation stage would have lasted for less than 10 seconds during the impact that formed the martian crater shown in Figure 4.27, and approximately 60–90 seconds for a larger crater, such as the Manicouagan crater shown in Figure 4.28.

- ■ In what ways is the impact process so different from most other geological processes?
- ❑ It occurs over much shorter timescales (seconds and minutes, rather than millions of years) and involves much higher pressures than any other geological process that exposes rocks at the surface of a planet.

Modification

Once the expanding shock waves have weakened and moved beyond the crater rim, they play no further part in crater formation. After this point, modification of the transient cavity depends on gravity and the mechanical strength of the target. Small craters (Figure 4.27) preserve the approximate original shape of the transient cavity, with only minor modification as debris cascades down the walls to form a layer of **brecciated** material in the base of the crater (Figure 4.30e and f).

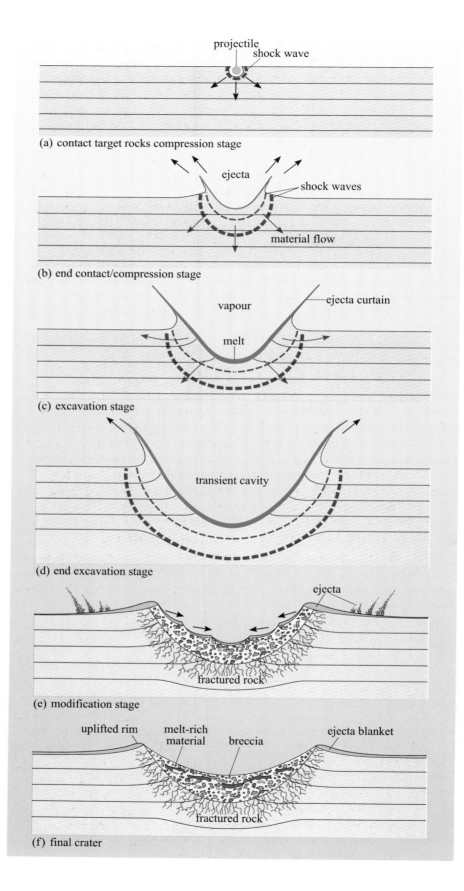

Figure 4.30 Schematic diagram portraying the development of a simple impact structure through the stages of contact and compression, excavation, and modification. Black arrows show the directions in which target material moves, blue arrows show the directions in which shock waves move at different stages of crater formation.

In larger craters, the transient cavity cannot sustain itself and it frequently collapses under gravity. Terraces may form due to landslides and slumping on the walls, as rocks at the edge of the transient cavity collapse inwards. Some craters may contain significant volumes of glassy material, formed from target rock that melted during the impact. As the floor of the crater beneath the impact structure begins to rebound, a central peak forms. This uplift is about one-tenth the final diameter of the crater so, for example, rocks beneath a crater that is 100 km in diameter will be uplifted vertically by 10 km during the impact event's modification stage (Figure 4.31). In still larger craters, under certain conditions, concentric mountain ranges may appear around the central impact. Over much longer timescales, as the crust of the planet gradually accommodates the impact, the crater may flatten out until it is defined by little more than slight differences in the material's appearance.

When molten rock is cooled (quenched) very quickly it forms a glass.

Even for large craters, 200–300 km in diameter, the modification stage will be completed 15 minutes after impact. To appreciate the scale of the impact process associated with a large impact consider that within 15 minutes an area of the Earth as flat as the East Anglian fens in the UK, or some of the desert areas of the USA or Australia could be transformed into a terrain having as much vertical relief as some of the Earth's major mountain ranges.

Additional effects

In addition to the physical excavation and modification of an impact crater, a variety of other processes occur during an impact. Large impactors may create a hole in the atmosphere of the target planet following their passage through it. In fact, for large objects as the impactor strikes the surface the trailing edge is still on the edge of space. Since the impact process happens so quickly the space vacated by the rapidly disappearing impactor effectively becomes a vacuum. Vapour and debris from the impact may be drawn upwards into this partial vacuum to high levels in the atmosphere. In addition, ejecta from the impact – glass, and unmelted rock fragments – may be accelerated at high speeds to form an apron of material around the impact site. Some material may be ejected at sufficient speed to completely escape a planet's atmosphere. Glass that re-enters the planet's atmosphere may fall thousands of kilometres from the impact site, with samples exhibiting aerodynamic shapes. Finally, ejecta from an impact may be accelerated with sufficient speed to escape a planet's gravity, leading to an exchange of material between planets.

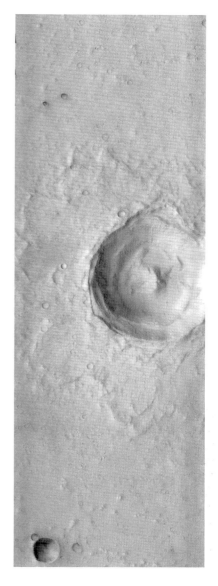

Figure 4.31 This Mars Odyssey image shows an example of a martian impact crater with a central peak. Central peaks are common in large, fresh craters on both Mars and the Moon and are formed during the extremely high-energy impact cratering event. In many martian craters the central peak has been either eroded or buried by later sedimentary processes, so the presence of a peak in this crater indicates that the crater is relatively young and has experienced little degradation. The inner walls of this approximately 18 km diameter crater show complex slumping that occurred during the modification stage of the impact event. Since that time there has been some downslope movement of material to form the small gullies that can be seen on the inner crater wall.

QUESTION 4.5

Which one of the following statements about the transient cavity produced in the impact process is *correct*?

A The diameter of the transient cavity will be smaller than the diameter of the projectile.

B The diameter of the transient cavity will be larger than the diameter of the projectile.

C The diameter of the transient cavity will be the same as the diameter of the impactor.

4.3.3 Types of impact craters

The specific shape a crater will have, i.e. its morphology, depends on a large number of factors including the size and speed of the impactor, the composition of the impactor and target rock, the strength and porosity of the impactor, the angle of impact, and the gravity of the target planet. However, a few broad classes of crater morphology may be recognized.

Simple craters

A typical example of a simple crater on Earth is Meteor Crater in Arizona (Figure 4.32). It is similar to others found on all solid surfaces in the Solar System. Simple craters are up to several kilometres in diameter, and are bowl-shaped depressions that lack a central uplift or terracing (Figure 4.33).

> The rim-to-floor depth of simple craters is approximately one-fifth of their diameter.

Complex craters

These craters are common in the Solar System (Figure 4.34). The transition between simple and complex craters depends largely on the gravity of the target body and the strength of the target material. On Earth the transition point between simple and complex craters may be as low as 2 km in soft sediments (4 km in harder rocks), while on the Moon the transition point occurs in craters between 10 km and 20 km in diameter. Mars shows the transition at about 7 km.

> The transition between simple and complex craters varies inversely with the target body's strength of gravity – the higher the strength of gravity the smaller the diameter that marks the distinction between the two.

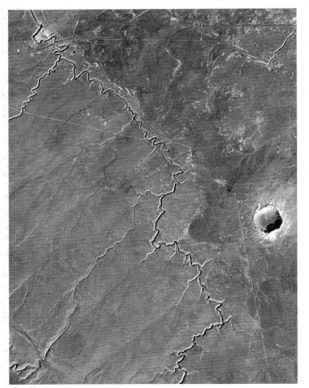

Figure 4.32 Meteor Crater is a 1.3 km diameter, 174 m deep simple crater in the flat-lying desert sandstones in Arizona. This image was acquired by the Landsat 4 satellite.

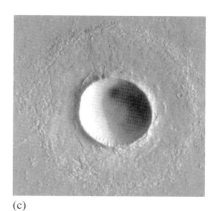

Figure 4.33 Simple craters: (a) and (b) on the Moon; (c) on Mars. The lunar craters are 7 km and 3 km in diameter, respectively; the martian crater is 2 km in diameter. Note the similarity with Meteor Crater (Figure 4.32). Apart from debris infilling their bases, these bowl-shaped depressions retain the approximate shape of the transient cavity.

Figure 4.34 (a) The 93 km diameter lunar crater Copernicus. (b) The 100 km diameter Peridier crater (26° N, 84° E) on Mars as imaged by Mars Global Surveyor. The dark feature on the floor of the crater is known from high-resolution images to be a field of windblown dunes. (c) A Mars Odyssey image of an un-named complex crater in the Arabia Terra area of Mars.

Complex craters are characterized by terraces of slumped blocks of rock, with the terrace width decreasing inwards, and central peaks formed by the rebound of rock during the impact.

The images in Figure 4.34 show several of the main features characteristic of impact craters that distinguish them clearly from volcanic craters:
- the crater floor is lower than the ground-level beyond the crater
- the crater is surrounded by a blanket of ejected material
- the inner wall of the crater has slumped into a series of terraces.

Elongate craters

It transpires that almost all craters are circular because the impact process results in an instantaneous release of energy (it is effectively an explosion). As such, the impact angle is largely unimportant until, that is, the impact angle is below about 10° at which point elongate craters may be produced. One such feature is the huge Orcus Patera structure on Mars, which is 380 km × 130 km and may be evidence of a low-angle impact (Figure 4.35). In extreme low-angle impacts, the resulting features cease to be true explosion structures as the object does not penetrate the target but ricochets away. If Orcus Patera is a low-angle impact structure, the object that formed it would have been huge, possibly greater than 20 km in diameter.

The elongate shape of Orcus Patera (14° N, 179° E) is clearly visible on the *Topographic Map of Mars*.

- ■ What else could Orcus Patera be other than an impact crater?
- ❑ Patera refers to the shape, i.e. an irregular crater often with scalloped edges (Table 1.1). On Mars most patera features are interpreted as volcanic in origin. Orcus Patera was given its name based on shape; its precise origin is still open to debate.

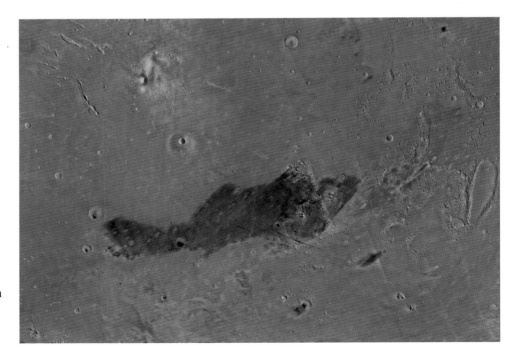

Figure 4.35 Orcus Patera crater (14° N, 179° E) in the Elysium region of Mars. This crater, 380 km in length, may be a low-angle impact structure.

Multi-ring basins

These are the largest impact structures that have been observed: none are known on Earth, but Valhalla Basin on Callisto (Figure 4.36), a satellite of Jupiter, is the largest known impact structure in the Solar System. On Mars the 1800 km diameter Hellas Planitia (Figure 4.37) is a possible multi-ring impact basin. The complex system of scarps that surrounds the central basin of multi-ring basins makes it difficult to define the diameter of the actual crater, but the overall features are huge:

- Valhalla has scarps encircling the basin that are up to 4000 km in diameter
- the 1800 km basin at the centre of Hellas Planitia is approximately 9 km deep and is encircled by a ring of material that rises 2 km above the surroundings and stretches out to 4000 km from the basin centre (Figure 4.38).

Multi-ring basins remain poorly understood, and it may be that these features are not a simple progression from complex craters but in fact have partially disrupted the entire tectonic framework of the bodies on which they are observed.

Figure 4.36 Voyager image of the vast Valhalla impact site on Callisto. Valhalla is the largest known impact structure in the Solar System, with a diameter of 4000 km, yet its topography is so subdued that each of the numerous visible ring scarps rises only 1–2 km above its surroundings.

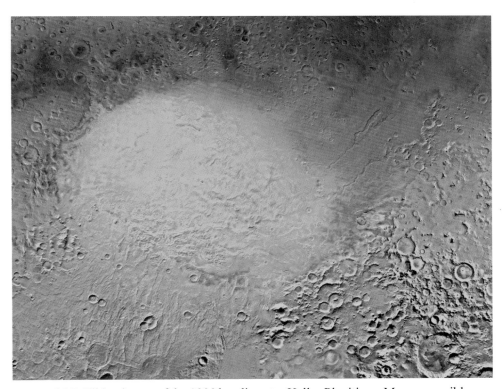

Figure 4.37 Viking image of the 1800 km diameter Hellas Planitia on Mars, a possible multi-ring impact basin and the largest impact structure on Mars. The exact diameter of Hellas Planitia is difficult to determine because large portions of the rim are missing to the northeast and southwest. In addition several large patera volcanoes occur along or near the rim and their flows have partially buried the older impact deposits.

Figure 4.38 Topography of Hellas Planitia on Mars obtained from laser altimeter data from the Mars Global Surveyor spacecraft. It is 9 km deep and 1800 km across, the basin is surrounded by a ring of material that rises to 2 km and stretches out to 4000 km from the basin centre. Blue represents low ground relative to the Mars datum, red high ground. Outside the rim of the crater are several large inward-facing escarpments that could be remnants of multiple rings.

4.3.4 Impactors and targets

Types of impactor

- ■ Are all impacting objects likely to be of the same composition?
- ❏ No, impactors fall into two broad populations: asteroids and comets.

Although the size and speed of the impactor may be the principal determinant of the size and morphology of the final crater, the composition and density of the object are also important. Apart from having distinct compositions, comets and asteroids also have substantial differences in relative speeds, density and source regions.

Asteroids are relatively dense, composed of silicate rock and metal, and most originate in the asteroid belt between Mars and Jupiter. They are deflected into the inner Solar System following collisions in the asteroid belt and gravitational interactions with Jupiter. At 1.52 AU (the orbit of Mars) they have average impact velocities of 11 km s^{-1}.

- ■ How does the density and source region of a comet differ from an asteroid?
- ❏ Comets are low-density objects made of ices and minor amounts of silicates, which originate in the Kuiper Belt near the orbit of Pluto, or the Oort cloud (see Chapter 1).

Occasionally a Kuiper Belt body may be disturbed by the interactions of the giant planets so that its orbit crosses that of Neptune. A subsequent close encounter with Neptune will change its orbit again, and may send it into the inner Solar System where it might be visible from Earth as a comet. Bodies residing in the Oort cloud are much too far out to be disturbed by planets, but they are in fact so far out that their orbits are affected by other stars! Again, some of these may fall towards the inner Solar System and be observed as comets. Given the different source regions for these populations, it is clear that the inner Solar System is subject to impacts from both comets and asteroids, whilst the outer planets rarely experience asteroidal impacts. Instead, planets and satellites in the outer Solar System may be subject to a much higher number of comet impacts.

Nature of the target

On Earth, the comparatively small number of distinct craters is testimony to our geologically active planet where renewing and wearing away of the surface occurs on a relatively short timescale when compared to the age of the Solar System. Abundant craters on other planets, satellites and asteroids suggest much less geological activity. In some cases, information on the geology of the target can be derived from the types of craters observed.

Figure 4.39 Fluidized ejecta apron around Arandas (42° N, 345° E), a martian crater 28 km in diameter. One prominent lobe of 'flow' material extends out from the crater rim to about one crater diameter. Two or three other series of lobes are present, one reaching more than 50 km from the crater rim.

Many martian craters look as though they were formed by objects that impacted into surfaces that had the consistency of wet cement (Figure 4.39). It is thought that the distinctive ejecta patterns of these **rampart craters** result from the mobilization of trapped groundwater, or melting of permafrost ice on impact, yielding ejecta with different properties from those formed on 'dry' planets. The observation of **fluidized ejecta** suggests the presence of abundant volatiles, such as carbon dioxide or **water-ice**, in the subsurface.

Craters can also provide information on the specific target rock type. The observation of impact craters in dunes on Mars presents the most unambiguous evidence that Mars has sedimentary rocks. If the dunes were made from loose sand, the craters would be rapidly eroded. Instead, it appears that the sand in these martian dunes was cemented prior to impact into a form of sandstone (Figure 4.40).

Figure 4.40 Mars Global Surveyor image of dunes on Mars showing impact cratering. Craters would not be supported in unconsolidated sand, so these images constitute excellent evidence for the existence of lithified (i.e. well consolidated) sedimentary rocks on Mars. This image is just over 2 km across.

4.3.5 Using craters to estimate the age of a planet's surface

In the lead-up to the Apollo Moon landings of the late 1960s and early 1970s, attempts were made to image the surface of the Moon in greater and greater detail. Each refinement in technology revealed more craters, until images from the first spacecraft to land showed that the Moon's surface is cratered down to a scale of millimetres. It became apparent that the longer a surface has been exposed to impacts, the more craters it will exhibit.

> The numbers of craters on a surface can be used to estimate its age: older surfaces have been exposed to impacts for longer and show more craters.

Complicating factors

The concept of using the numbers of craters to estimate the surface ages is a relatively simple one to grasp, but there are a number important questions that we should ask so that we can get a meaningful surface age. To start with, depending on whether we measure the number of asteroids that we can observe astronomically, or look at the number of craters on a surface of a known age, we need to scale from the diameter of the impactor to the diameter of crater that the object would produce.

- ■ Are all impactors of the same size?
- ❏ No, there is an enormous range in size.

We therefore need to know how the size of impactors varies, from the smallest dust to the largest asteroids. Photographs of the Moon show that for every large crater there are many smaller ones. By taking the measurements of all the crater sizes we would then be able to describe the relative numbers of objects across a range of sizes.

- ■ Has the rate of impacts remained constant throughout the history of the Solar System?
- ❏ No, the rate was much higher very early on.

Estimating the impact rate and the size distribution of impacting bodies is difficult, but these estimations are made even more difficult by the fact that both may vary in time, and throughout the Solar System. The impact rate was extremely high during the first few 100 million years or so of the Solar System as the planets finished accreting and probably peaked around 3.8–4.0 billion years ago, producing the so-called late heavy bombardment that is recorded in the large lunar impact basins. Since approximately 3.8 billion years ago, the impact rate has declined and is now at a low rate. Mars experiences a higher rate of impacts due to its proximity to the asteroid belt.

- ■ Do the properties of the target remain constant over time?
- ❏ No, the Earth's surface is very dynamic there have also been significant changes on Mars.

We therefore have to consider those properties of the target that will influence cratering statistics over time. For instance, billions of years ago Mars probably had a much denser atmosphere than it does today, which may have shielded it from small impactors in the same way that the Earth's atmosphere shields us today. We might expect a deficit of ancient small craters in this case. In addition, surface processes, which may have been more intense on Mars during the first 1–2 billion years of its history, may have aided crater degradation.

- ■ What will happen if a surface is bombarded continuously by impactors?
- ❏ If projectiles are continuously fired at a surface, early formed craters will eventually be obliterated by younger ones.

When a new crater can form only at the expense of an older one, by overprinting it, the surface is **saturated**. Obviously, the pre-saturation history of such a surface is irretrievably lost. Crater-counting statistics become unreliable if the origin of the craters is dubious. Large impacts produce huge numbers of **secondary craters**, so problems can arise if these are included within the population of primary impacts. Statistical techniques have been devised to cope with this issue, such as not counting small craters, many of which may be secondary craters.

- ■ Will counting the number of craters on a surface give us an absolute age for that surface? (For example an age equivalent to that we can obtain by radiometric dating.)
- ❏ No. Crater counting by itself offers at best only a means of comparing *relative* ages of surfaces.

However, if we have an absolute age for a surface, and then count the craters on it, we can produce a calibrated cratering curve. Because samples returned from the Moon have been dated by radiometric techniques, as have cratered terrains on Earth, we can use crater statistics in an *absolute* time frame for these bodies. Comparative crater statistics can then be used to estimate reliably the ages of lunar surfaces that have not been dated in the laboratory. But this confidence does not extend far, and so, based on knowledge of the Earth and Moon, and observed impactors, we have to make assumptions about the rate at which impacts have occurred in other parts of the Solar System if we want to use crater numbers to estimate the ages of surfaces.

Using craters to estimate the age of martian surfaces

Because no surfaces of Mars have yet been sampled for radiometric dating, crater statistics are generally used to estimate the age of its surface features. Excellent images of Mars are now available thanks to spacecraft like Mars Global Surveyor, so the crater numbers are themselves reliable. However, before these crater statistics can be used as a means of estimating the age for martian geological processes they have to be calibrated. This is done by comparing the data for Mars with that of the Moon for which we have samples of rock returned by the Apollo missions that have been radiometrically dated.

■ Why would the Earth be an unreliable source of data for calibrating martian cratering statistics?

❏ The Earth's surface is dynamic and craters are rapidly destroyed. Any statistics would therefore be unreliable.

As Mars has a larger mass than the Moon, and therefore stronger gravity, it experiences a higher rate of impacts over a period of time. Mars is also much closer to the orbiting 'scrapyard' of the asteroid belt, so there are more potential impactors. Both these factors indicate that Mars experienced a larger impact rate than the Moon and both have to be taken into account when age estimates of the martian surface are being produced.

4.4 Martian fluvial and aeolian processes

On Earth, the effects of water and wind – **fluvial** and **aeolian** processes, respectively – are major surface modification processes. However, these effects are rarely observed elsewhere in the Solar System. Mars is a notable exception, and on this planet we see abundant fluvial and aeolian landforms.

Since the Viking missions of the mid-1970s, it has been apparent that Mars has experienced volcanic, aeolian and fluvial activity that has substantially modified its surface. However, it was generally believed that this took place several hundred million years ago, or even billions of years ago. High-resolution pictures from the Mars Global Surveyor mission have changed this view. As you have seen, volcanic activity has occurred relatively recently on Mars, and there is cratering evidence for the existence of sedimentary rocks and perhaps subsurface water-ice. Additional images from the Mars Global Surveyor spacecraft have provided a wealth of information on past and current martian surface processes.

4.4.1 Fluvial processes – the action of water

The Viking missions revealed that Mars has experienced a variety of fluvial processes. There is evidence of:

- huge flash floods, extending over hundreds of kilometres (Figure 4.41a)
- branching river-beds, similar to terrestrial drainage systems, suggesting more mature, long-lived river systems (Figure 4.41b).

However, it is clear from crater-counting evidence that neither type of surface drainage has been active for at least 2 billion years. In an extremely exciting development over recent years, Mars Global Surveyor images have revealed a third type of martian drainage feature – small channels, located on the sides of older major valleys or on crater walls (Figure 4.42). These channels are too small to be visible in Viking images. Although dry (no liquid water has been seen on Mars), they appear to be very recent. Many of these surfaces show no discernible craters, and so fall below the limit of resolution of crater counting for determining surface age. This implies that they were made in the last 1 million years, which is extremely recent on a geological timescale, suggesting that fluvial processes are still occurring on Mars. We'll look at the evidence for water on Mars, and its importance to the search for life, in the next chapter.

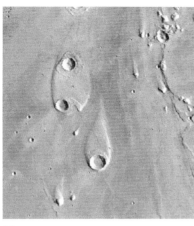
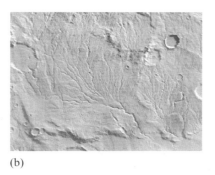

(b)

(a)

Figure 4.41 Evidence for massive flash floods in Mars' distant past. (a) Viking image showing teardrop-shaped craters, formed as floodwaters flowed around them. The craters are 8–10 km in diameter, lying near the mouth of Ares Vallis (10° N, 334° E) in Chryse Planitia. The height of the scarp surrounding the upper island is about 400 m, while the scarp surrounding the southern island is about 600 m above the plain. The region pictured is close to the Mars Pathfinder landing site (19° N, 326° E). Branching drainage systems (b) are evidence for more sustained drainage, similar to terrestrial river systems. This type of feature appears to be confined to older terrains, suggesting that sustained river systems on Mars have not been active for several billion years.

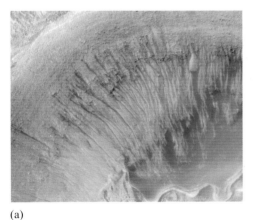

(a)

(b)

Figure 4.42 Mars Global Surveyor images showing evidence of recent drainage. Many craters have small channels in the crater walls (a), which appear to be unaffected by subsequent cratering, and have sharp sides, suggesting very recent formation ages. The image is about 5.5 km across. (b) Dunes showing very fresh channels emerging from close to the dune crest. This dune field in Russell Crater (55° S, 12° E) near the martian equator is frost covered in the martian winter so it is possible that the channels have been active during a recent spring thaw. The field of view is 4 km across.

4.4.2 Aeolian processes – the action of wind

When Mariner 9 arrived at Mars in 1971 it found the planet obscured by a global dust storm. Such storms are common on Mars (Figure 4.43), so it is not surprising that there are a variety of aeolian landforms, with wind-blown material forming dunes with a range of sizes. A huge variety of dune types may form, depending on the nature and availability of material, and the local wind conditions. Figure 4.44 shows several types of dune that have been observed on Mars.

The images provided by Mars Global Surveyor have shown that Mars remains an active planet. There is evidence of recent aeolian activity – new deposition and erosion have taken place in a single martian year – as well as mobility of volatile material like carbon dioxide in the martian polar regions (Figure 4.45).

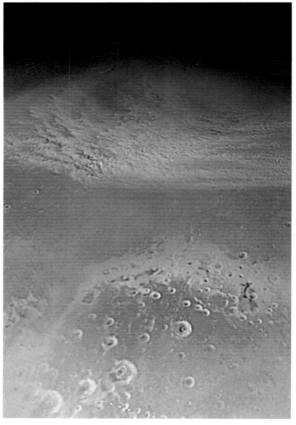

Figure 4.43 As Autumn got underway in the northern hemisphere of Mars in May 2003, Mars Global Surveyor witnessed large dust storms beginning to form on the northern plains bringing colder air down from the north polar cap. This early autumn view shows an eastward-moving dust storm on the plains north of Cydonia and western Arabia Terra. The storm is nearly as big as the continental United States are wide, from west to east.

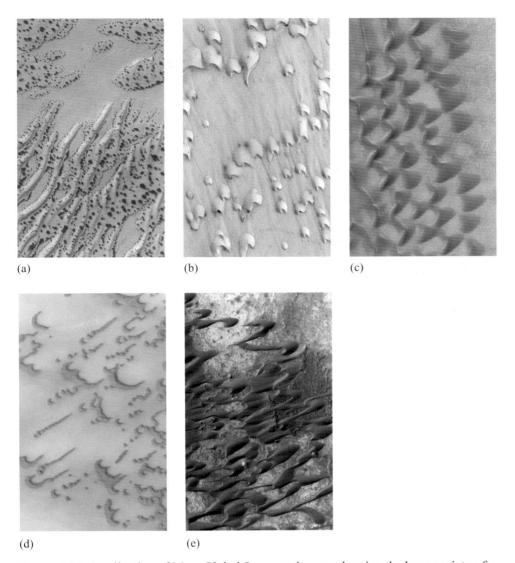

Figure 4.44 A collection of Mars Global Surveyor images showing the huge variety of martian dune forms. The dunes shown in (a) are in the middle of a spring thaw, with dark patches of ice-free sand just beginning to appear. The dunes in (b) are also frost covered – in summer they appear almost black – and are found in Chasma Boreale (83° N, 313° E), a giant trough that almost cuts the north polar cap of Mars in two. The shapes of the dunes suggest that wind is transporting sand from the top of the image towards the bottom. (c) Dunes and smaller ripples. (d) North polar dunes, the morphologies here suggest a reduced sediment supply. (e) A field of dunes in a volcanic depression in northern Syrtis Major. The images in a, c, d and e are 3 km across, while that in b is 1.5 km across.

Figure 4.45 Evidence of continuing surface processes on Mars; each image is approximately 250 m across. (a) and (b) Different areas of the martian south polar cap, the pictures in (a) taken in 1999, and (b) taken in 2001. When you compare the same area in different years it is clear that small hills have vanished and pit walls expanded between 1999 and 2001. The pits are formed in frozen carbon dioxide, and a little more sublimates away each year. (c) and (d) are the same portion of a ridged terrain north of the Olympus Mons volcano. The dark streaks are thought to result from avalanching of fine dust. The differences between these two images have occurred over the course of a single martian year.

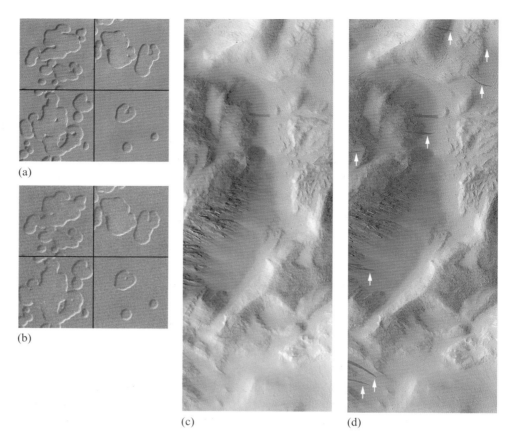

4.4.3 Sedimentary rocks

On Earth, fluvial and aeolian processes are fundamental parts of a geological cycle that includes the formation of sedimentary rocks. Eroded material, typically deposited in basins following weathering of upland regions, is cemented by percolating fluids to form a sedimentary rock. As we have seen, cratering of fossilized dunes provides excellent evidence for the presence of sedimentary rocks on Mars. In addition, Mars Global Surveyor images provide numerous examples of beautifully layered terrains. Figure 4.46 shows layering in a 64 km-wide impact crater.

Figure 4.46 Different areas of a martian layered terrain in an impact crater in the western Arabia Terra region. Each image is about 3 km across. Hundreds of layers of similar thickness and texture have been revealed by erosion, to produce the thin parallel lines that are visible in these images. This impact crater was a site of repeated sedimentary deposition, possibly related to cyclic changes in climate. It is not known whether the sediments were deposited dry or settled out of water that may have occupied the crater as a lake.

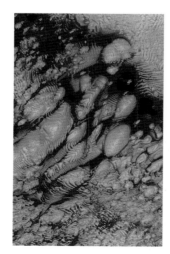 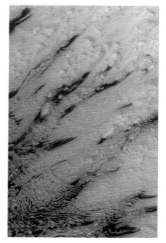 

Figure 4.47 Layered sedimentary rock exposed within the Becquerel Crater (22° N, 352° E). The image shows hundreds of light-toned layers in the 167 km wide crater. These layers are interpreted to be sedimentary rocks deposited in the crater at some time in the distant past. Sets of parallel geological faults can be seen cutting across the layers in the left third of the image.

The layers drape over the crater walls, forming beds at an angle near the walls. The layers are nearly horizontal in the middle of the crater, suggesting that the material was deposited from above. Towards the end of its initial mission mapping the planet, Mars Global Surveyor acquired one of its most spectacular images of layered sedimentary strata on Mars in the Becquerel crater (Figure 4.47).

4.5 Summary of Chapter 4

- The heat retained or generated within a terrestrial-like planetary body represents a key control in shaping the nature of the planetary surface.
- Primordial heat was produced in accretion and collision processes operating in the early stages of planetary evolution, and represents one of the important heat sources within terrestrial-like bodies. The other two important sources are radiogenic heating and tidal heating.
- Volcanism is the surface expression of melting processes occurring within a planet. It describes the processes associated with the transfer of magma, volatiles, and suspended crystallized material from interior to surface and, if present, into the atmosphere.
- There are two main types of volcanic activity: effusive volcanism, which is characterized by the production of lava flows, and explosive volcanism, which is characterized by the production of eruption columns and pyroclastic materials.
- Effusion of basaltic lavas has been the most widespread manifestation of volcanism on Mars.
- Impact cratering is the most pervasive process affecting the surfaces of solid bodies in the Solar System and unless decelerated by an atmosphere, impacts between different objects in the Solar System occur at very high speeds, typically more than 10 km s^{-1}.
- The formation of an impact crater may be divided into three distinct phases: contact and compression, excavation, and modification. The first two phases only last for a few seconds; the modification phase lasts only for a few minutes. Even in the largest impacts, cratering is the most rapid geological process known.
- Craters are observed in a wide range of sizes, from microscopic to thousands of kilometres in diameter. Simple craters of a few kilometres across are bowl-shaped depressions, larger complex craters may have central peaks and terraces, and even larger multi-ring basins may partially disrupt a planet's entire tectonic structure.

- There is a variety of impactor types. In the inner Solar System, impactors may be either asteroidal debris derived from the asteroid belt or comets from the Kuiper Belt and the Oort cloud.
- Impact craters provide a means of dating many planetary surfaces.
- There is abundant evidence for the action of water on Mars in the distant past, for example areas that show the effects of flash floods, or the dry beds of braided river systems similar to those found on Earth. Recent data suggests that small channels may still be active.
- Aeolian processes are ongoing. Changes in the surface of Mars have been observed over times as short as a single martian year.
- Sedimentary rocks, with detailed, repeated layering have been observed on Mars; some of these rocks may have been laid down in water.

4.6 Questions

QUESTION 4.6

What features distinguish an impact crater from a volcanic crater?

QUESTION 4.7

List the main characteristics of martian volcanoes that distinguish them from volcanoes on Earth.

QUESTION 4.8

Which *one* of the following statements about planetary heating is *incorrect*?

A Heat from the Earth's lithosphere is gradually lost to space over time.

B A primordial source of heat on Earth is heat generated by the decay of radioactive elements in the planet's interior.

C Mars is likely to have generated more modest amounts of internal heat than Earth.

D Mars has lost some of its internal heat by advection.

E Rocks can flow when subjected to high pressures and temperatures.

QUESTION 4.9

The shape of which *two* of the following surface features on Mars would suggest they are most likely to be volcanoes? Hint: refer to Table 1.1.

A Beer Crater

B Medusae Fossae

C Diacria Patera

D Alba Catena

E Jovis Tholus

QUESTION 4.10

In five or six sentences, summarize the main stages of the impact process.

QUESTION 4.11

Which of the following processes could be responsible for layered deposits on Mars?

A   Fluvial processes

B   Aeolian processes

C   Impact processes

D   Volcanic processes

E   All of these processes

QUESTION 4.12

Table 4.2 is designed to allow you to compare the significance of surface-modifying processes on the Earth and Mars.

(a) Complete the table by listing the main processes you've met in this chapter that form and modify surface features on the Earth and Mars in the column marked 'Process'.

(b) Give each process a ranking that gives an estimate of that process's relevance for Earth and Mars.

Rankings: 0 = absent, 1 = low, 2 = medium, 3 = high

Enter the rankings for Earth and Mars in the respective columns.

Table 4.2 For use with Question 4.12.

Process	Earth	Mars

CHAPTER 5
MARS AND LIFE

5.1 What is life?

5.1.1 Introduction

Imagine you are a visitor with no preconceptions approaching the Earth from a distant planet. As you move closer your view of the planet improves and you are able to see more and more detail. Does this planet support life? How can you decide? Before we can consider the question of life on Mars, therefore, we must ask some fundamental questions: What do we mean by life? What conditions does life, as we understand it on Earth, need in order to exist? What makes a planet habitable? What can we learn from the study of extreme environments on Earth? These questions will form the basis of our study of the present and past environment on Mars. This being a science course, we will examine these issues from a scientific viewpoint rather than a philosophical or religious one.

The science of life's origins and the possibility of life elsewhere in the Universe is known as Astrobiology.

QUESTION 5.1

As a thought experiment, spend a few minutes trying to come up with a definition of life based on your own general knowledge.

In attempting to define life, you have probably discovered just how unexpectedly difficult it is to perform this seemingly simple task. Don't worry, you're in good company, as some of the world's most famous scientists have struggled with the same difficulties in attempting to define what we mean by life. Yet it is only in the last 200 years that scientific questions have been raised about how and when life arose. Prior to this time, from a scientific perspective life was generally considered to have arisen spontaneously (Box 5.1).

BOX 5.1 SPONTANEOUS GENERATION OF LIFE

Prior to around 200 years ago, most scientists and philosophers believed that life arose spontaneously and repeatedly. It was a conviction that was apparently supported by evidence, for example the daily appearance of flies and maggots on rotting meat. Occasionally the idea of spontaneous generation was queried. A Tuscan doctor called Francesco Redi demonstrated in 1668 that maggots were the larvae of flies and if the meat was kept in a sealed container, so that adult flies were excluded, no maggots appeared. However, when Dutch microscope-maker Anthony van Leeuwenhoek detected micro-organisms in 1676, spontaneous generation was the seemingly obvious explanation for such ubiquitous creatures. The matter was finally laid to rest in 1862 when, in an attempt to win a prize offered by the French Academy of Science, Louis Pasteur (Figure 5.1) performed a convincing series of experiments. Pasteur showed that if a broth or solution was properly sterilized and excluded from contact with micro-organisms, it remained sterile indefinitely.

Figure 5.1 Louis Pasteur, who disproved the idea of spontaneous generation of life.

5.1.2 A definition of life

If we are to consider whether life exists or may have existed on Mars, we must first define exactly what we mean by life.

Most biologists would identify two key features that indicate life:

- the capacity for self-replication
- the capacity to undergo **Darwinian evolution**.

For an organism to self-replicate it must be able to produce copies of itself. For Darwinian evolution to occur, imperfections or mutations must occasionally arise during the copying process and these new mutations must be subjected to natural selection (Box 5.2). Nature favours particular characteristics under particular environmental conditions and those individuals able to exploit changing conditions are most likely to survive. Evolution occurs when any advantageous features brought about by mutation are passed on to further copies.

> From the list of two characteristics, a working definition of life can be created, namely: a self-sustaining chemical system capable of undergoing Darwinian evolution.

However, any such definition of life is likely to fail in certain circumstances. For example, the mule is an offspring of a donkey and a horse. Since mules are generally sterile they cannot breed and are therefore incapable of taking part in the processes of self-replication and Darwinian evolution, yet few would deny that it is alive. But, for the majority of cases, our definition of life will be a satisfactory one.

For life to be self-sustaining and capable of Darwinian evolution both energy and materials must be extracted from the surrounding environment to allow growth and replication. Furthermore, some sort of living apparatus must be present to govern and facilitate the chemistry of life.

BOX 5.2 NATURAL SELECTION AND DARWINIAN EVOLUTION

Figure 5.2 Charles Darwin.

After graduating from Cambridge with a degree in theology at the age of 22, Charles Darwin (Figure 5.2) set sail as a naturalist on the British Navy's HMS *Beagle* mapping expedition (1831–1836). His voyage took him to the Galápagos Islands in the Eastern Pacific Ocean. It was there that he began to formulate ideas about the process of evolution. Darwin recognized that any population consists of individuals who are all slightly different from one another. Those individuals having a variation that gives them an advantage in staying alive long enough to reproduce successfully are the ones that pass on their traits more frequently to the next generation. Subsequently, their traits become more common and the population evolves. Darwin called this 'descent with modification'.

The Galápagos finches provide an excellent example of this process. For instance, among the birds that ended up in arid environments, the ones with beaks that were better suited for eating cactus seeds got more food. As a result, they were in better condition to mate. In a very real sense, nature selected the best-adapted varieties to survive and to reproduce. Darwin called this process 'natural selection'.

Darwin did not believe that the environment was producing the variation within the finch populations. He correctly thought that the variation already existed and that nature just selected for the most suitable beak shape and against others. Darwin described this process as the 'survival of the fittest'. It was not until he was 50 years old, in 1859, that Darwin finally published his theory of evolution in a book entitled *The Origin of Species by Means of Natural Selection*. Today the concept of natural selection and its influence on successive generations is called Darwinian evolution.

5.1.3 The importance of carbon and water

There is only one chemical element that can form **molecules** of sufficient size to perform some of the functions necessary for life as we understand it on Earth. This element is carbon and all currently known life utilizes carbon-based organic compounds. You will often encounter the term '**organic**' in relation to how life may have originated. To some people 'organic' signifies the influence of biology, but to the chemist the term simply covers the chemistry of compounds based on carbon bound to the elements hydrogen, nitrogen, sulfur and phosphorous.

> Carbon can form chemical bonds with many other atoms, allowing a great deal of chemical versatility. It can also form compounds that readily dissolve in liquid water and, as you will see shortly, water is essential for life on Earth.

Chemical bonds are the mechanism by which atoms are held together to form molecules.

The relative abundance of the more common elements in the Universe indicates that the Universe is well stocked with the elements needed to construct organic compounds (Table 5.1). If we ignore the noble gases helium and neon, the four most common elements that are utilized by life on Earth (hydrogen, oxygen, carbon and nitrogen) are also the most abundant elements in the Universe. Sulfur and phosphorus (not listed, but the 15th most abundant element in the Universe) are also important for life on Earth.

Noble gases are highly unreactive and, until recently, were known as the inert gases. They include helium, neon and argon. These gases do not usually bind chemically with any other elements to form compounds.

Table 5.1 The ten most abundant elements in the Universe and in a human being as a pertinent example of life (expressed as a percentage of all elements).

Universe		Humans	
hydrogen	92.7%	hydrogen	60.6%
helium	7.2%	oxygen	25.7%
oxygen	0.05%	carbon	10.7%
neon	0.02%	nitrogen	2.4%
nitrogen	0.02%	calcium	0.23%
carbon	0.008%	phosphorous	0.13%
silicon	0.002%	sulfur	0.13%
magnesium	0.002%	sodium	0.08%
iron	0.001%	potassium	0.04%
sulfur	0.001%	chlorine	0.03%

A solvent is a liquid that can dissolve another substance or substances to form a solution.

Elements fundamental to the development of living organisms must be able to interact with one another, and that occurs most readily in the presence of water. Water has been called the universal solvent because it performs this function so well. Few other solvents can match the abilities of water to facilitate life. Water exists as a liquid in a temperature range that is not too cold to sustain biochemical reactions and not too hot to stop many organic bonds from forming.

> Two properties distinguish the Earth from other planets in our Solar System and enable it to support abundant life on its surface: liquid water covers much of its surface; the planetary environment maintains the water in its liquid state.

Indeed, it is the ability of water to exist as a liquid that is a major criterion used in defining whether or not a planet falls within a star's **habitable zone** (Box 5.3). Ammonia would be liquid on other worlds much colder than ours, but at such low temperatures chemical reactions that could lead to life would operate sluggishly and living systems may struggle to become established.

BOX 5.3 HABITABLE ZONES

On the Earth's surface pure water exists as a liquid between 0 °C and 100 °C. The presence of liquid water on a planetary surface could be used as a simple requirement for us to consider the planet as being habitable. In reality, this single factor is unlikely to be either necessary or sufficient, but it provides a useful guide for the conditions needed to support Earth-type life on terrestrial planets in orbit around a star. The primary consideration in determining a planet's habitability is therefore its surface temperature and a habitable zone is defined as encompassing the range of distances from a star for which liquid water can exist on a planetary surface.

The temperature of the surface of a planet in our Solar System is determined by the balance between the amount of energy it receives from the Sun and the amount of energy it loses back to space. In the case of the Earth this balance, combined with the slight warming due to the greenhouse effect of the atmosphere, is sufficient for water to exist as a liquid on its surface. However, the amount of energy coming from the Sun has not been constant over the history of the Solar System, but has slowly increased. Around 4 billion years ago, the energy from the Sun is estimated to have been around 70% of its present value.

- What effect will the slowly increasing amount of energy coming from the Sun over time have on the Sun's habitable zone?
- ❏ The habitable zone will migrate away from the Sun.

One consequence of the increase in energy coming from the Sun is that there has been a region around the Sun that has remained habitable throughout the history of the Solar System known as the continuously habitable zone (Figure 5.3).

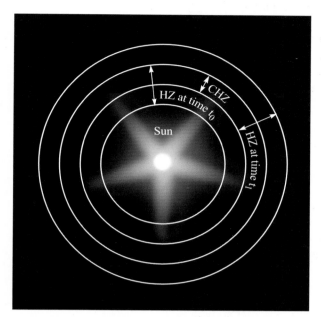

Figure 5.3 The habitable zone (HZ) around the Sun will move outwards as the energy coming from the Sun increases over time. The region that remains habitable is the continuously habitable zone (CHZ). $t_0 = 4.5$ billion years ago; $t_1 =$ today.

Scientists now use complex computer models to determine the extent of our Sun's habitable zone over time. These models, which take account of a range of different parameters in determining whether water could exist as a liquid on or near a planet's surface, place the Sun's present habitable zone at between 0.95 AU and a maximum outer extent of 1.67 AU.

- Where does the orbit of Mars fall in relation to the present habitable zone?

- Mars, with a mean distance from the Sun of 1.52 AU, falls within the habitable zone, which extends from 0.95 AU to 1.67 AU. Water may have existed on Mars (Figure 5.4).

However, these same computer models place the extent of the habitable zone that has been continuously habitable over the last 4.6 billion years at between 0.95 AU and 1.15 AU. This places Mars outside the region where liquid water could have existed on its surface during the early history of the Solar System.

The concept of a habitable zone outlined here is a working hypothesis. It is very much in line with thinking about Earth-like planets and the kind of life that is supported on Earth. However, the concept begins to break down when examined in detail. For example, you saw in Section 1.2.1 that Europa, one of the satellites of Jupiter, may have liquid water beneath its icy surface. Many scientists believe that Europa may be capable of supporting some form of life, yet it is well outside the Sun's habitable zone.

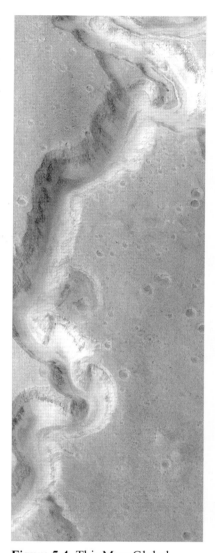

Figure 5.4 This Mars Global Surveyor image shows a portion of the meandering canyons of the Nanedi Valles system on Mars. The valley is about 2.5 km wide. The floor of the valley in the upper-right corner of the image exhibits a small channel, 200 m in width, which is covered by dunes and debris elsewhere on the valley floor. The presence of this channel is interpreted as indicating that the valley might have been carved by water that flowed through this system for an extended period of time.

Water molecules (H₂O) are the major component of living tissues, generally accounting for 70% of their mass.

- What clues are there in Table 5.1 to suggest there are relatively large quantities of water present in living systems?
- Hydrogen and oxygen, the elements that combine to form water, are the two most abundant elements in the human body.

Because living systems contain so much water it is inevitable that the majority of the other constituents necessary for life will exist in an aqueous environment.

5.1.4 The stuff of life

You've seen that life on Earth relies mainly on four elements – hydrogen, oxygen, carbon and nitrogen – together with smaller amounts of two other elements, sulfur and phosphorus. These six elements are found in a wide variety of organic combinations, and each combination has its own role in maintaining and perpetuating living systems. To fully appreciate how living systems operate we must begin to think of these elements in terms of the molecules in which they are contained.

Table 5.2 lists the major constituents that are found in a bacterium (a simple, single-celled organism). Note that the dominant constituent is water, but other chemical compounds are present as well.

Table 5.2 Types and abundances of the molecules that make up a bacterium.

	Percentage of total weight of bacterium	Number of types of molecule
water	70	1
inorganic material (e.g. sodium, potassium and calcium)	1	20
small organic molecules – typically containing less than 50 carbon atoms (e.g. amino acids, sugars, fats)	7	750
large organic molecules – typically containing more than 50 carbon atoms (e.g. lipids, carbohydrates, proteins, nucleic acids)	22	5000

- Are the majority of organic molecules in a living system present as small or large structures?
- Most of the organic molecules are very large.

Except for water, most of the molecules in a living system are large organic molecules. These can be subdivided into four different types: **lipids**, **carbohydrates**, **proteins** and **nucleic acids**.

These large molecules are usually the products of combining many individual organic molecules called **monomers** (from the Greek for single-parts) to create **polymers** (from the Greek for many-parts). Each of these types of polymer has a specific function in living systems. We'll briefly examine these different types of molecule and the roles that they perform.

Lipids (fats and oils)

Lipids are a diverse group of organic compounds, occurring in living organisms, which are insoluble in water but soluble in organic solvents. They include compounds such as fatty acids, oils, waxes and steroids, and have a wide range of functions in living organisms. Fats and oils are a convenient means of storing food energy in plants and animals. Lipids are able to group together in a flexible manner that enables them to play a critical role in cell membranes, which form part of the outer limit of cells.

Carbohydrates

Carbohydrates are a group of organic compounds that contain carbon, hydrogen and oxygen, and perform many vital roles in living organisms. The simplest carbohydrates are the sugars, including glucose and sucrose, which are essential intermediates in the conversion of food to energy. Larger, more complex carbohydrate molecules such as starch and cellulose are known as polysaccharides. Starches serve as energy stores for plants, for example in seeds; cellulose is the main constituent of plant cell walls and of vegetable fibres such as cotton.

The sugar used as a sweetener in everyday life is sucrose.

Proteins

Proteins (from the Greek *proteios* or primary) are the most complex large molecules found in living systems. They are manufactured by the joining together of a number of simpler molecules, known as amino acids, into compounds that have complex shapes. The sequence of amino acids determines the shape and this in turn determines the protein's function. Proteins are perhaps the most important of life's chemicals and have an enormous number of different roles. For example, they provide structure (e.g. in human fingernails and hair) and act as catalysts (e.g. aiding digestion in our stomachs). Proteins with catalytic properties are called **enzymes**.

A catalyst is a substance that increases the rate of a reaction but is not itself used up in the reaction.

Nucleic acids

Nucleic acids are the largest of the biological molecules and exist as a collection of individual **nucleotides** linked together in long linear polymers. Individual nucleotides contain:

- a sugar molecule
- molecules containing phosphorous and oxygen, known as phosphate groups
- a nitrogen-containing molecule known as a nitrogen-containing base.

The most famous nucleic acid is deoxyribonucleic acid or **DNA**. Prior to 1953 it was known that DNA contained four different nucleotides, each possessing identical sugar and phosphate groups but different nitrogen-containing bases. These nitrogen-containing bases are adenine, guanine, cytosine and thymine, often simply referred to by the first letters of their names: A, G, C and T. However, exactly how these components were arranged was unknown until 1953, when James Watson and Francis Crick recognized that DNA consists of two long molecular strands that coil about each other to form a **double helix** (Figure 5.5). The two strands are joined by chemical bonds that resemble the steps of a spiral staircase. The steps consist of two nucleotides, with each nucleotide forming half of the step. Weak bonds join the bases in the centre of the helix. The bases always match – adenine in one nucleotide is always paired with thymine in the other and, likewise, guanine is always linked to cytosine. Hence, the sequence of bases on one strand strictly determines the sequence of bases on the other.

DNA strand	DNA strand
A	T
T	A
G	C
C	G

The bases are attached to their helical strands by sugar groups, which in turn are connected together along the exterior of the helix by phosphate groups.

- If one strand of DNA has bases arranged in the sequence T-A-A-C-A-G, what will be the sequence in the other corresponding strand of DNA?

- Since A always bonds with T, T with A, G with C and C with G, the corresponding sequence must be A-T-T-G-T-C.

- Can you suggest how this complementary system allows DNA to pass on genetic information?

- Since the sequence of bases on one strand of the helix determines the sequence on the other, 'unzipping' the double helix provides two templates that can be used to produce two new DNA molecules from the single parent.

Special protein enzymes that separate the strands of the double helix achieve the unzipping of DNA. The single strands hook up with spare nucleotide molecules present in the liquid surrounding the molecule. Each base in the unzipped strand latches on to its complementary base. The nucleotides then join together in helical strands; so two identical double-helix molecules are formed, exactly like the original (Figure 5.6).

> The discovery of the structure of DNA provided the basis for understanding one of the key characteristics of life: the mechanism that enables biological molecules to replicate themselves.

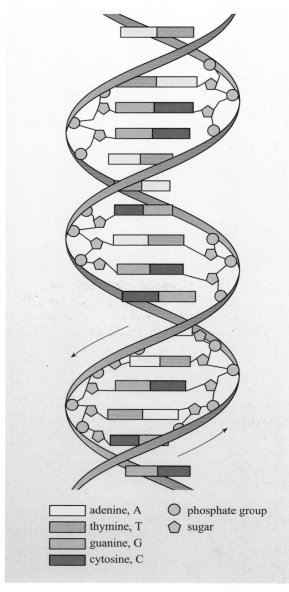

Figure 5.5 The DNA double helix. Note that the ribbons are not real, but are there to illustrate the nature of the double helix.

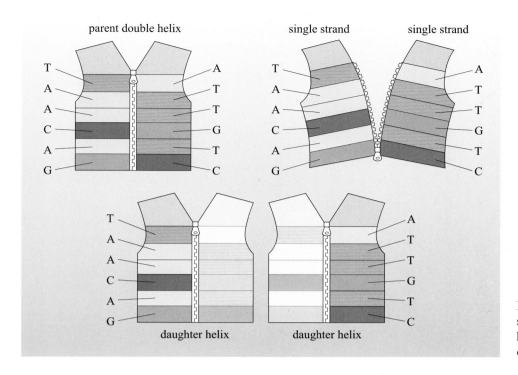

Figure 5.6 DNA replication showing how one parent double helix produces two identical daughter double helices.

In addition to self-replication and passing on information from generation to generation, DNA also governs the production of proteins in an organism. DNA contains a set of 'instructions' called the **genetic code** that is expressed by the sequence of bases in the molecule. For example A-T-T-G would be one part of a genetic code, while A-T-G-G would be another. This genetic code directs the production of thousands of proteins needed for the structure and function of living systems and is achieved via another nucleic acid called ribonucleic acid or **RNA**. RNA is also the molecule that some viruses use to store their genetic information rather than DNA.

The cell

Many different molecules must be in close association for living systems to operate. This is because in chemistry the rate of a reaction generally increases with the concentration of the molecules reacting. Yet what is there to stop molecules simply drifting off in solution and bringing a halt to the chemistry of life? The answer is the cell. In its simplest form, a cell is a small bag of molecules that is separated from the outside world (Figure 5.7). At the centre of the cell, strands of DNA are devoted to the storage and use of genetic information. The DNA is surrounded by a salt water solution containing enzymes and other chemicals, known as the **cytosol**. A soft membrane, called the cell membrane, which is built of lipids and proteins, surrounds the cell contents. The cell membrane restricts the movement or flow of molecules into and out of the cell and thereby protects the cell's contents. Finally, in plants, algae, fungi and bacteria (but not animals) a tough cell wall consisting of carbohydrate molecules and short chains of amino acids provides the cell's rigidity. So cells provide an environment in which biochemical processes can occur and genetic information can be stored. Cells are the basic structural unit of all present-day organisms on the Earth, but vary in number, shape, size and function. For instance, bacteria are single-celled organisms whereas humans contain many billions of cells.

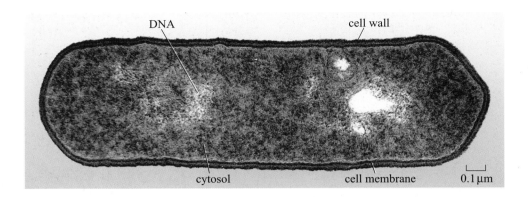

Figure 5.7 A simple cell.

Simple cells such as the one in Figure 5.7 can reproduce by splitting into two. This process begins when DNA unzips to form two template molecules; when the copying process is complete there are two new DNA molecules. These attach themselves to different parts of the cell membrane. Next the cell begins to divide, separating the two DNA-containing regions. Finally, when cell division is complete, two identical daughter cells have been produced from the parent.

5.1.5 Biomarkers: finding signatures of life

Now that we have explored what we mean by life, what is necessary for life to exist and what living organisms are made of, we can turn to the detection of life. There is more than one approach to the problem and scientists use the term 'biological marker', or 'biomarker' for short, for any evidence that indicates present or past life.

There are several categories of biomarker that might be relevant to the search for life on Mars:

- cellular remains
- textural fabrics in sediments that record the structure and/or function of biological communities, including fossils or the traces left by organisms (Figure 5.8)
- biologically produced organic matter
- minerals whose deposition has been affected by biological processes
- atmospheric constituents whose relative concentrations require a biological source.

A brief glance at these criteria will reveal the extreme subjectivity that arises when attempting to define a biomarker. Establishing whether or not textural fabrics or organic matter in a sample is biological in origin might be a very difficult process. For example, hydrocarbons are a class of organic molecule that can be generated by heat and pressure on the biological remains of living organisms and are a major component of coal. Coal comprises the fossil remains of land plants that were alive millions of years ago. However, exactly the same sorts of hydrocarbons can also be produced by internal combustion engines, garden barbecues or giant stars. It is easy to imagine the controversy that could arise from detecting such organic

Figure 5.8 Examples of some of the oldest fossils found on Earth, the 670 million year old Ediacaran fossils. (a) *Dickinsonia*, an elongate pancake-shaped worm; (b) *Cyclomedusa*, a jellyfish.

compounds and attributing their presence to once-living organisms. As you'll see in Section 5.3, such difficulties presented particular problems for the first attempts to detect evidence of life on Mars by the Viking missions in the 1970s.

5.1.6 The extremes of life

It is important that we examine how life on Earth responds to extreme environmental conditions. After all, what we might consider an extreme environment on Earth may be closer to normal conditions on Mars. Can life survive or even thrive under extremes of temperature and pressure? Can it exist in environments that lack water or that are exposed to high levels of radiation? The answer to both questions is a surprising 'yes'; indeed we have ample evidence of life's ability to tolerate extremes from a range of environments on Earth.

> The many micro-organisms that are capable of different degrees of adaptability to the extreme range of living conditions available on Earth are known as **extremophiles**. The number of known extreme-loving organisms from a variety of environments is increasing, confirming that life can exist, and indeed thrive, under those conditions.

The ability of life to survive in extreme environments has been dramatically demonstrated by the recovery of bacteria exposed to the hostile environment of the lunar surface. The astronauts of Apollo 16 recovered a camera from the Surveyor 3 spacecraft, which had landed on the Moon two and a half years earlier (Figure 5.9). When returned to Earth, foam from the inside of

Figure 5.9 Recovery of the camera from the Surveyor 3 spacecraft by the Apollo 16 astronaut, Pete Conrad. When returned to Earth, a strain of the bacterium *Streptococcus* that was isolated from foam inside the camera was found to have survived exposure on the lunar surface.

the camera was examined to see if any bacteria on it had survived their journey to the lunar surface. The 50–100 organisms recovered survived launch, space vacuum, three years of radiation exposure, deep-freeze at an average temperature of −250 °C, and no nutrient, water or energy source. However, the organisms did nothing while they sat on the lunar surface, they were in effect freeze-dried. In fact, despite surviving such a hostile environment, these bacteria could not be considered extremophiles in the true sense; they were merely able to survive in this hostile environment.

> An important observation about extremophiles is that these organisms do not merely tolerate their lot; they do best in their punishing habitats and, in many cases, require one or more extremes in order to reproduce at all.

So what are the physical limits to life on Earth and what sort of organisms thrive under extreme conditions? Extremes of temperature, radiation, desiccation and pressure are particularly relevant to the search for past or present habitats for life on Mars.

Temperature

Temperature presents a range of challenges to living organisms. The structural breakdown of cells caused by the formation of ice crystals in sensitive plants can be readily witnessed in those parts of the world that experience cold winters or even just the occasional frosty night. At the other extreme, high temperatures result in the structural breakdown of biological molecules such as proteins and nucleic acids. Temperatures of 100 °C can disrupt the structural integrity of cell membranes to the extent that they leak important cellular constituents.

Life on Earth has adapted to a surprising range of temperatures (Figure 5.10). Although the majority of organisms grow best at moderate temperatures of between 15 °C and 50 °C (the **mesophiles** in Figure 5.10), the temperature preferences of other organisms range from **hyperthermophiles** (able to reproduce at temperatures >80 °C) to **psychrophiles** where maximum growth occurs at temperatures <15 °C.

< is the symbol for less than; > is the symbol for more than.

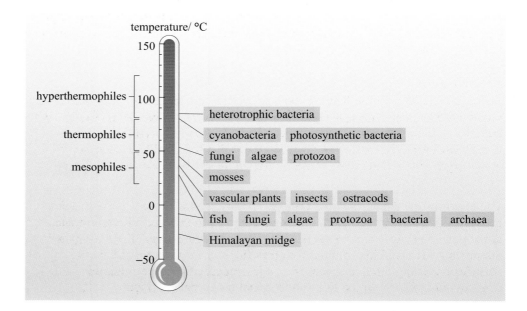

Figure 5.10 The temperatures at which major groups of organisms thrive.

No complex animals or plants are known that can tolerate temperatures above 50 °C for prolonged periods of time, though microbial **thermophiles** that are content at temperatures up to 60 °C have been known for a long time. True extremophiles, those able to flourish in greater heat, were first discovered in the 1960s during a study of microbial life in hot springs and other waters of Yellowstone National Park in the USA. To date, more than 50 species of hyperthermophiles have been isolated, the most resistant of which grow in the walls of black smokers (sea-floor geothermal vents) where they reproduce in an environment of about 105 °C; indeed many hyperthermophilic organisms won't grow at all at temperatures below 90 °C.

What is the upper temperature limit for life? Do 'super-hyperthermophiles' capable of growth at 200 °C or 300 °C exist? At present we do not know, although it seems likely that the upper limit is about 140 °C. Above this temperature, proteins and nucleic acids break down, so that a loss in the integrity of DNA and other essential molecules would probably prevent reproduction. So how have organisms adapted to these high temperatures? Thermophilic and hyperthermophilic organisms appear to have adapted by having DNA and proteins that are better able to cope with higher temperatures, and more stable cell membranes.

On Earth, cold environments are actually more common than hot environments. The Earth's oceans maintain an average temperature of 1–3 °C. However, large areas of the Earth's surface are permanently frozen or are unfrozen for only a few weeks in summer. Some of these frozen environments support life in the form of psychrophiles. Representatives of all major groups of organism are known from environments with temperatures just below 0 °C. Freezing in liquid nitrogen at a temperature of –196 °C can preserve many microbes successfully. However, the lowest recorded temperature for *active* microbial communities is substantially higher, at –18 °C. Liquid water is both a solvent for life and an important reactant or product in many biological processes. When water freezes the resulting ice crystals can rip cell membranes apart, and the freezing of water inside cells is almost invariably lethal. Two principal adaptations have evolved to deal with temperatures below the freezing point of water: the protection of cells by preventing ice formation or, if ice does form, protection of the cells during thawing. One way organisms prevent ice forming is to accumulate compounds that can depress the normal freezing point of water, i.e. accumulate a sort of anti-freeze. For example, high concentrations of such anti-freeze compounds can enable the survival of some animals to temperatures as low as –60 °C.

Radiation

Radiation is energy in the form of waves or particles, such as electromagnetic radiation (e.g. gamma rays, X-rays, ultraviolet radiation, visible light, or infrared radiation, see Box 2.2). Very high levels of radiation do not occur naturally on Earth. However, Mars lacks the protection afforded the Earth's surface by its atmosphere, and in particular the Earth's **ozone** layer absorbs significant amount of ultraviolet radiation.

The effects of radiation on living organisms have been well studied as a result of research on the use of radiation in medicine and on the consequences of human activity ranging from warfare to space travel. Indeed, the level of radiation on the martian surface resulting from cosmic rays (high energy atomic or subatomic particles travelling through space) is a major concern for space agencies contemplating future exploration of the planet by astronauts (Figure 5.11).

Figure 5.11 The present-day risk to human life from cosmic-ray radiation on the martian surface is a serious health concern for any future human exploration of Mars. This map was produced from data returned by the Mars Odyssey and Mars Global Surveyor spacecraft and shows estimates of the amount of cosmic radiation that reaches the martian surface. The colours on the map refer to the estimated average number of times per year each cell nucleus in the human body would be hit on Mars by a cosmic-ray particle. The range is generally from a moderate risk level (two hits, colour-coded green), to a high-risk level (eight hits, colour-coded red). (Low areas are black.)

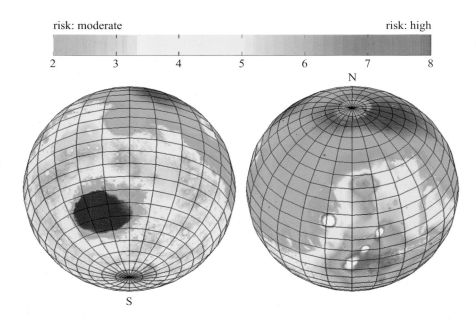

The Latin name of an organism uniquely identifies the species.

Radiation in the ultraviolet region of the electromagnetic spectrum can cause serious damage to DNA by causing breaks in the two strands that make up the DNA double helix. One organism that is known to withstand exceptional levels of radiation and that probably qualifies as a radiation extremophile is the bacterium *Deinococcus radiodurans*, first discovered in 1956. It can withstand exceptionally high doses of ultraviolet and gamma radiation. The ability to survive such extreme environments is attributed to the bacterium's ability to repair any damaged DNA. This extraordinary resistance is thought to be a consequence of evolutionary adaptations to cope with extreme desiccation, so that *Deinococcus radiodurans* is also a **xerophile** (see Table 5.3).

■ How does the risk posed by cosmic ray radiation shown in Figure 5.11 relate to the topography of Mars? *Hint*: compare Figure 5.11 with Figure 4.8 or with the *Topographic Map of Mars*.

❑ Radiation levels are highest in areas of high elevation, e.g. the southern highlands of Mars and lowest in areas of low elevation, e.g. the northern lowlands or the Hellas impact basin (the large round dark area in Figure 5.11).

■ What might be responsible for the apparent relationship between radiation levels and elevation on Mars?

❑ The lower areas have more atmosphere above them to block out some of the radiation. Earth's thick atmosphere shields us from most cosmic radiation, but Mars has a much thinner atmosphere than Earth does.

Desiccation

You saw in Section 5.1.2 that the high melting and boiling points of water and the wide range of temperatures over which it remains liquid, make water an essential solvent for life. Water limitation therefore represents a particularly extreme environment for life. Some organisms (xerophiles) can tolerate extreme desiccation by entering a state of apparent suspended animation, characterized by little water within their cells and a cessation of biological activity. This is well documented in organisms such as bacteria, yeasts, fungi, plants and animals associated with environments where the water essential for active life is often transient and sporadic. When the water disappears these organisms appear to be dead for periods of days, weeks, or even years until moisture returns, when they 'come back to life' and resume their normal activities.

Pressure

Terrestrial plants and animals at the Earth's surface have evolved at normal atmospheric pressure. However, pressure increases with depth in the oceans so marine organisms may have to deal with much higher pressures. Atmospheric pressure also decreases with altitude, so that by 10 km above sea-level, it is around one-quarter of the atmospheric pressure at sea-level. Pressure presents problems to life because it forces volume changes, for example when pressure increases, the molecules in cell membranes pack more tightly, restricting the flow of essential fluids through the cell membrane. Organisms that can tolerate high pressures have often adapted the compositions of their cell membranes to improve the flow of essential fluids and nutrients. Pressure-loving **piezophiles** have been recovered from the Earth's deepest sea floor, the Mariana Trench, where they thrive at pressures of more than 6000 times the pressure at the Earth's surface.

Atmospheric pressure on Mars is 0.006 times that of the Earth.

The strength of gravity also has an effect on the forces experienced by an organism. However, until recently, all organisms on Earth have lived at our normal strength of gravity. The advent of space exploration means that humans have had to deal with a range of different gravity regimes, from the huge forces experienced during launch to 'microgravity' environments on board the International Space Station (ISS). Although most research concerned with microgravity has concerned human health, studies on board the ISS have demonstrated that gravity plays an important role in a variety of biological processes. Scientists now believe that there are conditions in which the weightless environment influences the cellular machinery, resulting in specific changes to cell membranes and the reproduction of micro-organisms.

Microgravity is the very weak strength of gravity that occurs on board a spacecraft in orbit round the Earth.

pH and salinity

pH, which ranges on a scale from 0 to 14, is a measure of the acidity or alkalinity of a solution. Biological processes tend to occur towards the middle range of the pH scale so that typical environmental pH values also fall within this range, e.g. the pH of seawater is approximately 8.2. Some extremophiles are known that prefer highly acidic or alkaline conditions, the **acidophiles** and **alkaliphiles**. Acidophiles thrive in the rare acidic habitats having a pH of between 0.7 and 4, and alkaliphiles favour alkaline habitats with a pH between 8 and 12.5.

Highly acidic environments can occur naturally within rocks as a result of the passage of water through the rock, for example at some hot springs. However, acidophiles are not able to tolerate a significant increase in acidity inside their cells, where it would destroy important molecules such as DNA. Thus they survive by keeping the acid out. But the defensive molecules that provide this protection, as well as others that come into contact with the environment, must be able to operate in extreme acidity. Indeed, enzymes have been isolated from acidophiles that are able to work at a pH of less than 1 (vinegar has a pH of about 4).

Alkaliphiles live in highly alkaline soils and in so-called soda lakes, such as those found in Egypt, the Rift Valley of Africa and the western USA. Above a pH of 8 or so, certain molecules, notably those made of RNA, break down. Consequently, alkaliphiles, like acidophiles, maintain neutrality inside their cells.

Salts are typically water-soluble crystalline compounds containing either a metallic element (e.g. magnesium) or ammonium in combination with elements or groups of elements such as chlorine or nitrate.

Salinity is a measure of the total quantity of dissolved salts in water. Sodium chloride, which is common table salt, is one such salt but others are composed of elements such as magnesium and calcium. Organisms can live within a range of salinities, from distilled water to saturated salt solutions. **Halophiles** are organisms that require high concentrations of salt in order to live. Their optimal sodium chloride concentrations for growth range from twice to nearly five times the salt concentration of seawater. They are found in habitats like the Great Salt Lake (Figure 5.12), the Dead Sea and salterns (evaporation basins for obtaining salt). Some high-salinity environments are also extremely alkaline because weathering of sodium carbonate and certain other salts can produce alkaline solutions. Not surprisingly, microbes in those environments are adapted to both high alkalinity and high salinity.

Figure 5.12 The Great Salt Lake, Utah, seen from the Shuttle Discovery.

5.1.7 Extreme environments

The shear diversity of life on Earth makes it impossible to do a complete survey of the Earth's more extreme environments in a few pages.

> The continued discovery of new extreme environments and the organisms that inhabit them has made more plausible the search for life on Mars.

Given that many of these environments that appear to be extreme on Earth may be analogous to the normal environments on Mars, we'll briefly examine a few of them that may have a role to play in providing suitable habitats for life in otherwise hostile environments.

Hot springs

Hot springs and geysers are commonly associated with geothermal areas such as those of New Zealand (Figure 5.13). They are characterized by steam and hot water, and sometimes low pH and toxic metals such as mercury. They are, nonetheless, environments that sustain a remarkably diverse range of life. The range of colours visible in Figure 5.13 reflects different algal populations growing around the Waiotapu hot springs in New Zealand.

Geothermal hot springs occur where volcanic processes bring magma close to the surface and into contact with groundwater.

- Why might hot-spring environments be important in the search for evidence of past life on Mars?

☐ Mars has been volcanically active in the recent geological past (Section 4.2) so that heat from volcanism may have resulted in the production of environments on or near the surface of Mars where hot water may have existed.

- What sort of organisms might we find in hot-spring environments?

☐ The relatively high temperatures would favour thermophiles or hyperthermophiles, while the alkalinity of the water might favour alkaliphiles.

Figure 5.13 Hot spring in Waiotapu Park, Rotorua in North Island, New Zealand. The various colours around the edge are due to microbial mats formed by organisms that thrive in different temperature and pH environments.

Deserts

Deserts are environments on Earth that are extremely dry and can be either hot or cold. Water is always the limiting factor in such ecosystems. The Atacama Desert in Chile is one of the hottest and driest areas, while the coldest and driest places on Earth are the so-called dry valleys of Antarctica (Figure 5.14), where there is no ice cover, very little water and extremely low temperatures – almost Mars-like. The primary inhabitants of both kinds of desert ecosystems are bacteria, algae and fungi that live on or a few millimetres below the surfaces of rocks. Organisms that have adapted to living on, or beneath the surfaces of rocks are referred to as **endoliths** (from the Greek *endon*, meaning 'within' and *lithos*, 'stone').

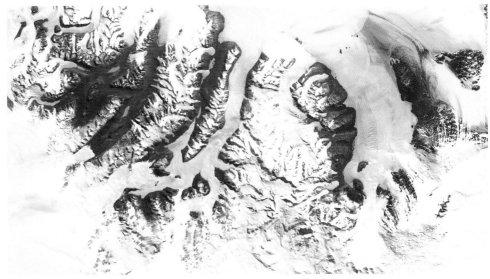

Figure 5.14 Satellite view of the McMurdo dry valleys in Antarctica, one of the most extreme environments on Earth. The dry valleys are large ice-free regions with average temperatures of around –20 °C. It never rains in the dry valleys, and only occasionally snows (equivalent to 10–20 mm of rainfall per year).

At first glance the dry valleys of Antarctica appear lifeless, yet endolithic organisms have been recovered there and the occasional presence of liquid water from melting glaciers in the summer months forms lakes that contain algae and bacteria.

Subsurface environments

As you will see in Section 5.2, it would be hard for life to survive the harsh conditions present today on the martian surface. However, one or two of the organisms we've examined in this section could hypothetically withstand one or more of the martian extremes, though they would need some protection. Mars, for the most part, is dry and frigid. It receives only 43% as much energy from the Sun as the Earth, but the thin carbon dioxide-rich atmosphere absorbs little of the harmful ultraviolet radiation.

- Of the various extremophiles covered in this section, which do you think might prove most resistant to the arid and exposed conditions on the martian surface?

- One of the toughest is *Deinococcus radiodurans*, which has evolved to cope with high radiation conditions and desiccation.

The search for present-day life on Mars is focused on the possibility of life existing below the surface.

The plausibility of subsurface life on other planets has been enhanced by the discovery on Earth of subsurface lithotrophic microbial ecosystems, abbreviated to SLiME (lithotrophic refers to an organism that obtains nutrients from rock). These microbes are not reliant on sunlight to survive. Instead, they appear to thrive on chemical energy in basalt, a rock common to Earth and Mars, but which contains little of the normal nutrients that feed micro-organisms. These microbes were found in groundwater samples taken more than 1000 m below the surface of the Columbia Basin basalt rocks in the western USA and appear to exist on a diet of mostly hydrogen. The basalt connection suggests that it might be possible for micro-organisms to exist in the martian subsurface. This does not mean there is life on Mars, but if such organisms can exist here then, in theory, they could also exist on Mars.

QUESTION 5.2

Which *one* of the following names would most accurately describe the kind of extremophile you might expect to find in an ice-covered Antarctic lake?

A xerophile
B mesophile
C psychrophile
D acidophile
E piezophile

Table 5.3 summarizes our present state of knowledge of the physical limits to life.

Table 5.3 Some of the known physical limits for life on Earth.

Environmental factor	Limiting conditions	Name	Comment
temperature	<15 °C	psychrophiles	Consist mainly of bacteria, algae and fungi. Psychrophiles have adaptations that enable then to survive low temperatures, e.g. cell membranes and enzymes that function optimally at relatively low temperatures.
	15–50 °C	mesophiles	The vast majority of organisms on Earth are mesophiles, inhabiting all the major temperate and tropical regions.
	50–80 °C	thermophiles	The majority are single-celled organisms found in hot springs and undersea hydrothermal vents. Thermophiles have adaptations that enable their cell machinery to function at high temperatures, including structural modifications of their proteins, nucleic acids and cell membranes to give them greater heat stability.
	80–121 °C	hyperthermophiles	Similar to thermophiles but able to tolerate even higher temperatures.
radiation			The bacterium *Deinococcus radiodurans* is the most radiation-tolerant organism known, having been recovered from irradiated materials and also from rocks from regions of Antarctica that are thought to most closely resemble martian surface conditions.
salinity	15–37.5% salt	halophiles	Bacteria that are able to grow in high concentrations of salt.
pH*	0.7–4	acidophiles	Organisms able to tolerate highly acidic environments.
	8–12.5	alkaliphiles	Organisms able to tolerate highly alkaline environments.
desiccation	dry conditions	xerophiles	A wide range of organisms are able to survive very dry conditions, including plants, fungi and bacteria.
pressure	high pressure	piezophiles	Organisms able to tolerate pressures hundreds of times that of atmospheric pressure at the Earth's surface. Some organisms may be able to tolerate the pressures produced by shock waves in meteorite impacts.

*pH is a numerical scale that runs from 0 to 14 and indicates the acidity or alkalinity of a solution. A solution is acidic if the pH is less than 7 and alkaline if greater than 7. Pure water has a pH of 7 and is neutral.

5.2 The martian environment

No aspect of Mars is more important in terms of possible life than the scientific understanding of the martian environment.

■ What two properties enable the Earth to support abundant life on its surface?

❏ Liquid water over much of its surface and an environment that maintains water in its liquid state (see Box 5.3).

In this section we'll examine the martian environment starting with the planet's atmosphere. We'll also discuss the martian climate and some of the weather systems that it produces before examining some of the current efforts to search for water on Mars.

5.2.1 The atmosphere of Mars

It has been known for a long time that Mars has an atmosphere. Drawings of telescopic observations from the 17th century onwards suggest the existence of the martian polar ice-caps. By the 1960s, when the first spacecraft missions flew by Mars, the presence of carbon dioxide and water had already been detected in the martian atmosphere. Nevertheless, little was known at that time of the detailed composition of the martian atmosphere and even less was known about its physical properties, such as temperature and pressure. Estimates of the atmospheric pressure on Mars, for example, were too high. Since the early 1960s our knowledge of the atmosphere of Mars has expanded enormously due to information from spacecraft and more sophisticated Earth- and space-based telescopes.

Atmospheric composition

Although Earth and Mars are both terrestrial planets, their atmospheres appear, at least on first inspection, to display more differences than similarities. The major constituents of each atmosphere are shown in Figure 5.15. The compositions given here are those at the surfaces, where the atmospheres are most dense. The interaction of energy from the Sun with an atmosphere at higher altitudes leads to chemical reactions that convert some of the molecules to different species. For example part of the oxygen component of the Earth's atmosphere is converted to ozone at higher altitudes. Consequently, the composition of the atmospheres changes with altitude.

Oxygen is a gaseous molecule composed of two oxygen atoms, whilst ozone is composed of three oxygen atoms.

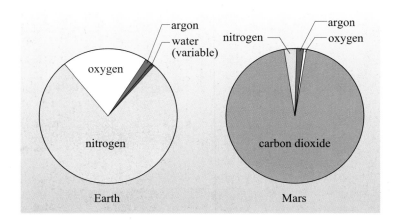

Figure 5.15 The major components of the atmospheres of Earth and Mars, as measured at the surfaces of the planets. The area of each slice of the pie chart is proportional to the amount of that gas in the atmosphere.

■ What is the major constituent of Mars' atmosphere?

❏ Carbon dioxide.

One major difference in the composition of the atmosphere of the Earth compared to that of Mars is the Earth's larger water content. The physical conditions on Earth ensure that most water exists in the oceans, which act as a reservoir. The Earth's polar ice-caps form another part of the water reservoir. On Mars, in addition to the atmospheric components, reservoirs of water and carbon dioxide are also contained in the polar ice-caps (Figures 5.16 and 5.17) and permafrost.

Permafrost is a term used to describe permanently frozen soil, subsoil or other deposits.

The northern polar cap of Mars is believed to comprise carbon dioxide overlying a residual cap of water-ice about 600 km across, which is exposed in summer. In contrast, the south polar cap appears to be composed predominantly of carbon dioxide.

So, not surprisingly, the Viking missions observed relatively large amounts of water in the atmosphere close to the north polar cap of Mars, especially during summer when the cap evaporates. Rather less enhancement of water was observed at the south polar cap in its summer.

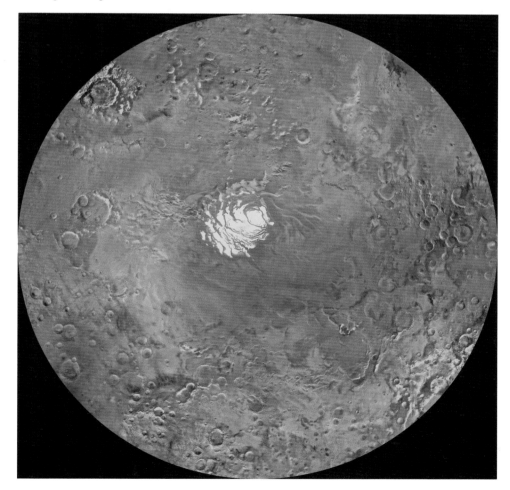

Figure 5.16 A view of the south pole of Mars obtained by the Viking spacecraft.

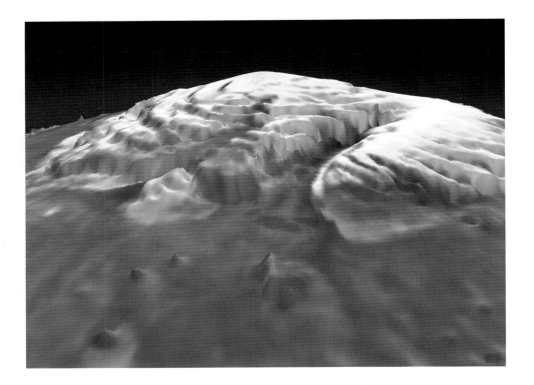

Figure 5.17 A 3-dimensional view of the north polar ice-cap of Mars. Scientists used laser altimeter data from the Mars Global Surveyor spacecraft to estimate the volume of water-ice in the north polar ice-cap. The image is vertically exaggerated to highlight the variations in topography of the region.

In contrast to the Earth, when the frost and ice of the martian polar caps are heated up they do not melt to produce liquid water but turn straight to a gas or vapour. Under most familiar conditions, if you heat a solid it will turn to liquid, and more heat needs to be applied to turn it to vapour. Melting frozen water and then heating it to its boiling point is the most familiar example. However, under appropriate conditions, solids can be transformed directly to vapour without melting first. 'Dry ice', the substance used on stage to generate clouds of mist, is frozen carbon dioxide, and at room temperature on Earth this turns directly to vapour without melting. This process is called sublimation, and the solid carbon dioxide is said to 'sublime' when it turns to vapour.

The factor that determines whether a solid will melt or sublime is the pressure of the vapour of the particular substance. Carbon dioxide is stable in liquid form only when confined by carbon dioxide gas (vapour) at a pressure of several times the Earth's total atmospheric pressure.

■ Can carbon dioxide exist as a liquid under the normal conditions on the martian surface?

❑ No, on Mars the pressure is far lower than the Earth's atmospheric pressure so liquid carbon dioxide cannot exist there.

Thus the frozen carbon dioxide in the polar caps sublimes directly to vapour when it is warmed by the Sun. As a consequence the size of the polar caps varies between summer and winter on Mars (Figure 5.18).

In order for water to be stable as a liquid, the pressure of water vapour in an atmosphere must exceed about one-hundredth of the Earth's total atmospheric pressure. This condition is not met today on Mars, so warming frozen water there

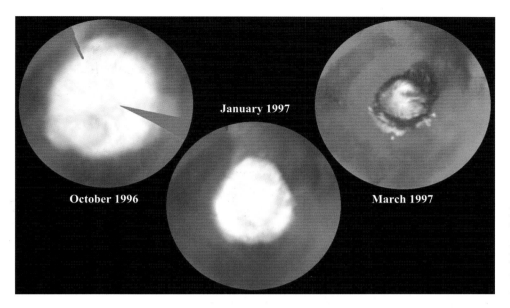

Figure 5.18 Hubble Space Telescope images of the changes in the northern polar ice-cap of Mars between martian early spring (October 1996) and martian early summer (March 1997).

causes it to sublime rather than melt. However, as you'll see in Section 5.2.3, there is evidence that Mars may have had a denser and more water-rich atmosphere at various times in the past, and this would have enabled liquid water to be stable under surface conditions.

■ What evidence have you already seen for the presence of liquid water at some time in Mars' past?

❑ Recall from Section 4.4.3 and Figure 4.47 that there is evidence of fluvial processes having operated on Mars.

One consequence of the sublimation of carbon dioxide and water on Mars is that there is a continual seasonal exchange of carbon dioxide and water between the polar caps via the atmosphere. (See Box 5.4 for an explanation of how seasons arise.) During summer in the martian northern hemisphere, carbon dioxide and water sublime and migrate to the southern pole where they condense as ices (Figure 5.19).

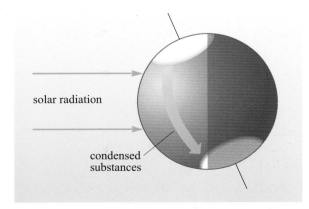

Figure 5.19 During summer in the martian northern hemisphere, energy from the Sun (solar radiation) warms the carbon dioxide and water-ice in the northern polar cap. These materials sublime and migrate to the southern pole where they form condensed substances.

BOX 5.4 SEASONS

Both Earth and Mars have similar length days, i.e. both planets spin around axes passing through their north and south poles in about 24 hours – the spin of a planet around its axis is called rotation. However, they take different lengths of time to complete one revolution of the Sun. The Earth takes 365 days and Mars takes 686.5 days to complete one full orbit of the Sun. In completing one revolution of the Sun the planets map out an imaginary plane, known as the orbital plane. Seasons are caused because the axis of rotation is tilted with respect to the orbital plane (Figure 5.20). The angle that the axis of rotation makes to a line drawn perpendicular to the orbital plane is called the axial inclination, i_a (Figure 5.20), and is about 24° for both Earth and Mars.

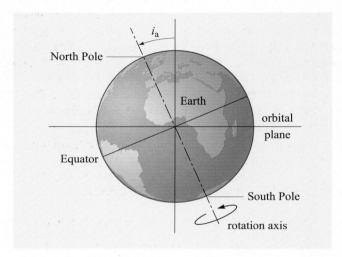

Figure 5.20 The axial inclination, i_a, of a planet.

When it is summer in the Northern Hemisphere on Earth, that hemisphere is tilted towards the Sun (Figure 5.21) with the result that day is longer than night and the Sun is high in the sky. At this time it is winter in the Southern Hemisphere and the opposite conditions prevail. When the Earth has completed a further half-orbit about the Sun, the seasons are reversed.

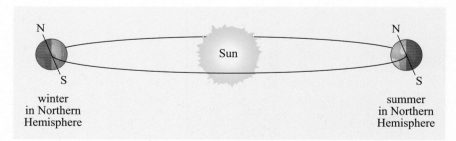

Figure 5.21 The seasons alternate as the Earth orbits the Sun (not to scale).

Atmospheric structure

On Earth, the temperature usually decreases with altitude, at least to the tops of the highest mountains. This variation of temperature with altitude is an aspect of **atmospheric structure**, and is determined largely by the absorption of energy from the Sun by a planet's atmosphere and surface. Together with pressure, which also decreases with altitude, temperature determines the formation of clouds in the atmosphere. The heating of the atmosphere and surface of a planet by solar radiation is also the cause of atmospheric motion.

Temperatures have been recorded at the surface of Mars and at different altitudes. Of course, these temperatures also vary with latitude, time of day and the season. To single out the variation with altitude, the temperatures are averaged over time and latitude. These variations of average temperature with altitude for Earth and Mars are shown in Figure 5.22. This figure indicates several similarities between the temperature variations in the atmospheres of both Earth and Mars. In the lowest region of each atmosphere, the temperature drops with increasing altitude. This region is called the **troposphere**, meaning the region of mixing. The rapid decrease of temperature with altitude, which is due to heating of the surface, causes convection, which leads to vertical mixing of the gases in the atmosphere. The uppermost region of each atmosphere, called the **thermosphere**, is characterized by an increase of temperature with altitude; this increase shows a large daily variation. Between these regions is the **mesosphere** in which the temperature decreases with altitude, but more slowly than in the troposphere. The **stratosphere**, which is a region between the troposphere and the mesosphere in which the temperature increases with altitude, is unique to the Earth.

■ How does the height of the troposphere on Mars compare with that on the Earth?

❏ On Mars the troposphere extends to almost 50 km, whereas on Earth it only extends to about 10 km high.

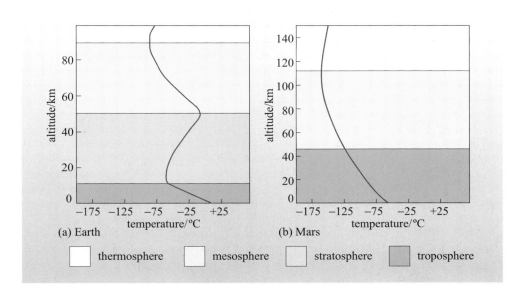

Figure 5.22 Atmospheric structure: the variation of temperature with altitude for (a) Earth and (b) Mars.

5.2.2 Mars climate and weather

The surface of Mars is rather cold. At low latitudes, a typical daily temperature range is from −100 °C to +17 °C, with a mean of about −60 °C. In winter, the temperatures at the poles can fall to −125 °C.

■ Can you think of a simple reason for the low temperature on the Martian surface?

❏ The most obvious reason is Mars' distance from the Sun, namely 1.52 AU.

A dramatic demonstration of the extremes of temperature of the martian environment was revealed in a study of night-time surface temperatures by the Mars Global Surveyor spacecraft and is shown on the map in Figure 5.23.

■ What are the lowest and highest night-time temperatures shown on the map?

❏ The lowest temperatures, shown in purple, are around −120 °C, while the warmest, shown in white, are around −65 °C.

The extreme range between day and night temperatures on Mars is one consequence of the thinness of the martian atmosphere. On the hemisphere of Mars facing the Sun, the surface heats up rapidly because the overlying atmosphere absorbs very little solar radiation. When this hemisphere has turned to face away from the Sun, radiation from the surface readily escapes from the planet, again because the atmosphere is so thin. The result is a large difference in temperature between day (temperatures up to +17 °C) and night (approximately −125 °C). This temperature difference causes a flow of atmosphere around the planet and is one of the major processes that affects the circulation of the martian atmosphere.

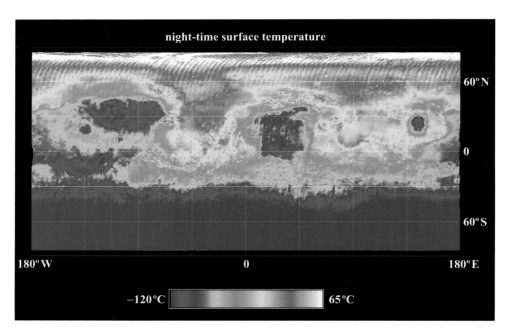

Figure 5.23 Night-time temperatures of rocks at the martian surface measured by the Mars Global Surveyor. The coldest temperatures are shown in purple, the warmest in white.

Atmospheric pressure of Mars (6 mbar compared to the Earth's 1000 mbar) varies by about plus or minus 15% as a result of the seasonal sublimation and condensation of carbon dioxide from the poles. Clouds, which are formed where a component such as water has condensed to form small liquid droplets or solid particles, do form on Mars (Figure 5.24) although the total average cloud cover is small. However, most clouds are composed of dust (Figures 5.24 and 5.25) raised by wind, and a combination of various atmospheric motions and local eddy currents (circular movements of the atmosphere) can lead to local surface wind speeds as high as 180 km/h (110 mph). These winds create local dust storms (Figure 5.25) or dust devils (Box 5.5). Within these storms the dust absorbs solar radiation, setting up large temperature differences and hence pressure differences. These lead to further turbulence causing more dust to be raised, which in turn heats up and leads to even more turbulence. In spring, these dust storms are strengthened by the higher pressure as the carbon dioxide sublimes, enabling more dust to be suspended for a longer time.

1 bar is atmospheric pressure on Earth at sea-level. 1 millibar (mbar) is one-hundredth of 1 bar.

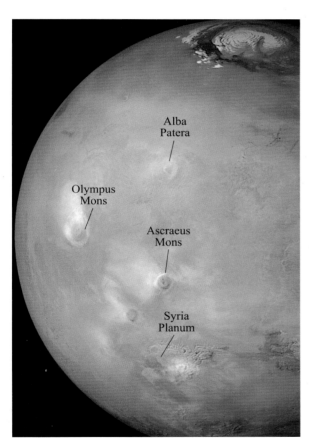

Figure 5.24 The weather on Mars. In this image acquired by Mars Global Surveyor a dust storm rages in Syria Planum (lower-centre of the image). Water-ice clouds are also present over each of the five largest Tharsis volcanoes, Olympus Mons (left centre), Alba Patera (upper centre), Ascraeus Mons (near centre), Pavonis Mons (towards the lower left), and Arsia Mons (lower left). The summertime north polar residual water ice-cap can be seen at the top of the image.

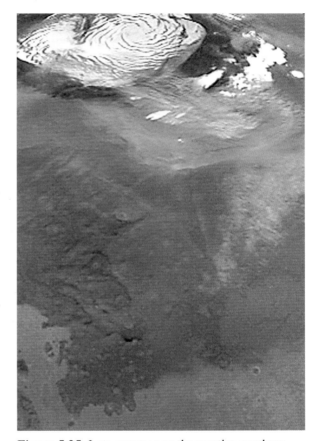

Figure 5.25 Late summer on the martian northern plains means clouds and dust storms. This image, acquired by Mars Global Surveyor on 7 March 2003, shows a dust storm near the north pole. Late-summer dust storms occurred nearly every day from late February until well into April 2003. The white features at the top of the image are the water-ice surfaces of the north polar ice-cap.

Over a martian year many local dust storms occur on Mars, and once or twice a year the whole planet becomes enveloped in a storm (Figure 5.26).

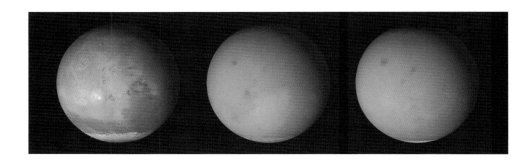

Figure 5.26 During June 2001 Mars Global Surveyor tracked a small dust storm into Hellas Planitia. Within a week the dust storm had enveloped the entire planet raising dust from separate locations in Arabia Terra and Hesperia Planum, thousands of kilometres away from Hellas.

BOX 5.5 MARTIAN DUST DEVILS

Dust devils are spinning columns of wind that move across the landscape, pick up dust, and look somewhat like miniature tornadoes. They are a common occurrence in dry and desert landscapes on Earth, as well as Mars. They form when the ground heats up during the day, warming the air immediately above the surface. As the warmed air nearest the surface begins to rise, it spins. The spinning column begins to move across the surface and picks up loose dust. The dust makes the vortex visible. On Earth, dust devils typically last for only a few minutes. Remarkably, they have been imaged on Mars by orbiting spacecraft (Figure 5.27).

When dust devils pass over the surface of Mars, they leave dark, criss-crossing streaks on the land. They can cross over hills, run straight across dunes and ripples, and go through fields of house-sized boulders. The dust devils remove bright dust from the terrain, revealing a darker surface underneath (Figure 5.28).

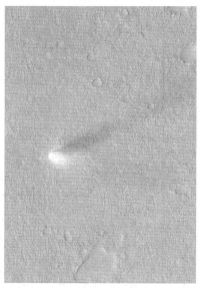

Figure 5.27 This Mars Global Surveyor image, acquired during the northern martian summer, shows a dust devil on the surface of Mars. Sunlight illuminates the scene from the lower left and the dust devil is casting a shadow towards the upper right. The view shown here is 3 km wide.

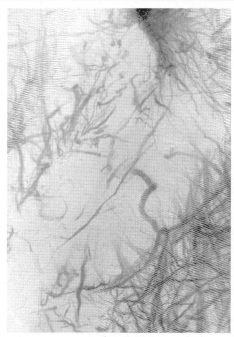

Figure 5.28 Mars Global Surveyor image of dust-devil streaks on the floor of Argyre Planitia (50° S, 316° E) on Mars.

QUESTION 5.3

Explain why the daytime and night-time temperatures on Mars can differ by more than 100 °C.

5.2.3 The search for water

Images of the surface of Mars, captured by the early space missions (Figure 5.29), showed features that resemble the canyons and valleys of Earth, whereas others look distinctly like river channels on Earth. These are all indications that water once flowed as a liquid on Mars. However, evidence from the martian cratering record suggested that these features were formed very early in the planet's history, between about 4 billion and 3 billion years ago.

As early as 1972, the first conclusive evidence for the existence of surface water came from images taken by the Mariner 9 spacecraft (see Figure 5.29a). However, the real progress in our knowledge of the existence of water on Mars has come with the results from the Mars Global Surveyor and Mars Odyssey spacecraft. These have produced an enormous body of evidence that reinforces the previously held idea that Mars once possessed significant amounts of water at or near its surface. Furthermore, the evidence points to water-related activity in relatively recent times and, more speculatively, at the present day. The image resolution of Mars Global Surveyor has pinpointed hundreds of delicately structured gully systems (Figure 5.30). Individual gullies are just 10 metres wide (earlier missions couldn't detect such small features because of the inferior resolution of the technology) and a whole system might cover an area of only a dozen football

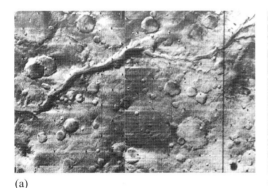
(a)

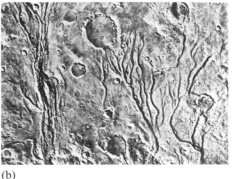
(b)

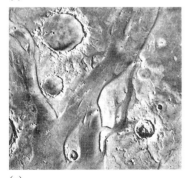
(c)

Figure 5.29 Images of the martian surface from early space missions, showing a variety of features indicative of flowing water. (a) Part of a channel, 700 km in length, in the heavily cratered southern highland region, discovered by Mariner 9 in 1972. Channels like these provide firm evidence of an episode of erosion by flowing water very early in Mars' history. (b) River valleys and impact craters. (c) A Viking Orbiter image showing the giant 'outflow channel' Ares Vallis (10° N, 334° E), a result of catastrophic flooding. The largest impact crater visible is 62 km in diameter. Streamlined islands with pointed prows upstream and long tapering tails downstream are clearly visible.

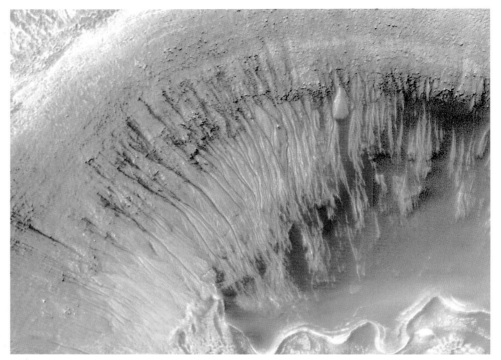

(a)

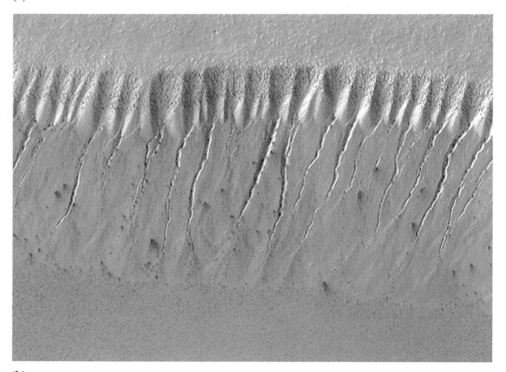

(b)

Figure 5.30 Two examples of martian gullies observed by the Mars Global Surveyor. (a) Gullies in the northern wall of the Newton crater (41° S, 202° E) in the northern hemisphere. The width of this image corresponds to a real distance of 6.5 km on Mars. (b) Gullies in the southern polar region (71° S, 4° E). The image covers an area of approximately 2.8 km wide by 2.1 km high.

pitches. Most gullies are in the southern hemisphere and their sculpted terrain, cut-bank patterns, and fan-shaped accumulations of debris look remarkably similar to flash-flood deposits in deserts on Earth (Figure 5.31). However, the starts of the gullies rarely have tributaries and are unlike systems fed by precipitation, so the cause appears not to be the same as flash floods on Earth.

Many (though not all) of the gully systems appear on the shaded sides of hills facing the polar ice-caps. Their geometry suggests that swimming-pool volumes of water could be entombed underground until suddenly it becomes warm enough for an ice plug to burst, letting all the water rush down the slopes. Such a scenario is shown in Figure 5.32. In this model, underground liquid water is trapped behind an ice barrier or 'plug' that has formed on the shadowed slopes of craters and ravines. Salts dissolved in the water behind the plug could help it to remain liquid as salts can lower the freezing point of water significantly. Ultimately, when the damn breaks, a flood is sent down the gully, resulting in the observed patterns.

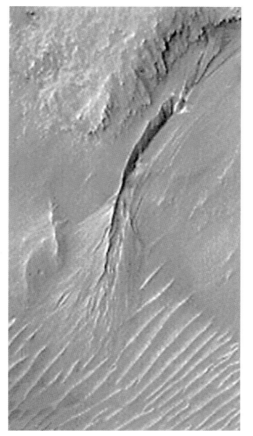

(a)

(b)

Figure 5.31 (a) The accumulated debris (or 'apron') from this gully in a south-facing wall in Nigral Vallis (29° S, 320° E) covers sand dunes that may have formed less than a century ago. The width of this image corresponds to a real distance of 1.6 km on Mars. (b) For comparison, an apron on Earth is shown. In this example, rainwater flowing under and seeping along the base of a recently deposited volcanic ash layer (at Mount St Helens) has created the gully.

Many of the gully systems look extraordinarily recent (less than 1 million years old) – they are sharply carved and cross older, wind-scoured features. Their appearance is so fresh, in fact, that some planetary geologists think that Mars may have undergone massive, short-term climate changes, where water could come and go in hundreds of years. Indeed, scientists wonder whether liquid water might exist on Mars today, buried in some areas perhaps 500 m underground.

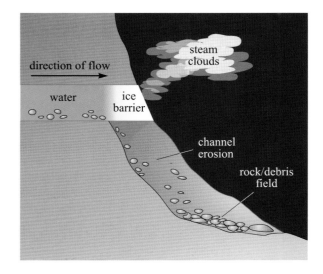

Figure 5.32 A possible model for the formation of the characteristic channels and aprons of martian gullies.

The findings are corroborated by data from another instrument on the spacecraft, the Mars Orbiter Laser Altimeter (MOLA). For 27 months – longer than a martian year – MOLA gauged the daily height of the polar ice-caps, meticulously recording how much frozen material accumulated in winter and sublimated in summer in each hemisphere (Box 5.4). MOLA showed that each ice-cap has a volume as great as the Greenland ice-cap on Earth.

MOLA and the Mars Orbiter Camera (MOC) also measured how the polar caps shrink in each hemisphere's summer. They found that there is enough water and carbon dioxide frozen in the polar caps that if one-third of it sublimated into the martian atmosphere it would pump the atmospheric pressure up from its present 6 mbar to 30–40 mbar (the Earth's atmospheric pressure is about 1000 mbar). Scientists believe that climatic changes on Mars may have been sufficient to achieve this relatively recently.

■ What effect will raising the amount of water vapour in the martian atmosphere have on the stability of water on the martian surface?

❑ At a pressure of 6 mbar, water is not stable as a liquid, it sublimes. By raising the atmospheric pressure and importantly the amount of water vapour in the martian atmosphere, water may become stable as a liquid under the right temperature conditions.

Figure 5.33 An artist's impression of Mars Odyssey's search for water on and below the martian surface. Mars Odyssey carries an instrument (a gamma-ray spectrometer) capable of mapping chemical elements on or near the surface of Mars.

Thus, perhaps in the recent past Mars might have been mild enough for ponds of water to have dotted its surface like desert oases.

One of the prime goals of the Mars Odyssey mission (see Section 2.6) is to search for evidence of water on the martian surface (Figure 5.33). This mission started its main scientific tasks at the end of February 2002 and in a very short time began to produce outstanding results. Mars Odyssey carries an instrument called the Gamma-Ray Spectrometer (GRS) that is capable from its orbit of detecting elements present in the martian soil. The instrument works by detecting gamma rays (Box 2.2) produced when cosmic rays (charged particles in space that come from the stars including our Sun) collide with the atoms of chemical elements in the martian surface. By making such measurements, it is possible to determine which elements are present, how abundant they are and how they are distributed around the planet's surface.

■ What are gamma rays?

❑ Gamma-rays are a form of electromagnetic radiation (Box 2.2).

■ What element could we look for to detect water on the martian surface?

❑ Water is H_2O, so either hydrogen or oxygen might be suitable.

Since most minerals also contain oxygen, Mars Odyssey's GRS has used the hydrogen signal to look for signs of water (Figure 5.34). As soon as the GRS started to take observations, it produced startling results that implied the existence of significant quantities of hydrogen. In fact, it gave indications of concentrations of hydrogen below the surface.

To date, the best explanation of the GRS data from Mars Odyssey is that the topmost layer of martian soil is hydrogen-poor and this overlies a layer that is hydrogen-rich. The data also suggest that the thickness of the upper hydrogen-poor layer decreases with decreasing distance to the pole. The hydrogen in the subsurface layer is generally believed to be in water-ice, and scientists have proposed a distribution of water-ice within the top couple of metres of martian soil (Figure 5.35). With these latest results from the Mars Odyssey GRS, we are at least partially answering the question, 'Where did all the water go that has at one time flowed on the surface of Mars?' But there is still a great deal to do. These results, for example, have only been able to comment on the situation in the topmost metre or so. We still need to know what is happening at greater depths and this is a major goal of future Mars missions.

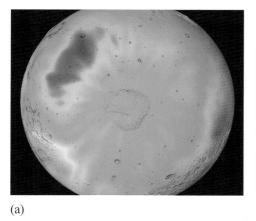
(a)

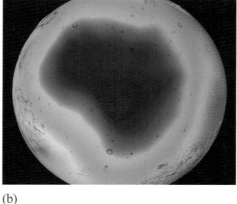
(b)

Figure 5.34 Distribution of hydrogen on or near the surface of the northern hemisphere of Mars in (a) martian winter and (b) martian summer. The maps are based on data from Mars Odyssey's gamma-ray spectrometer. Martian soil enriched by hydrogen is indicated by the purple and deep-blue colors on the maps. Progressively smaller amounts of hydrogen are shown in the colours light-blue, green, yellow and red. In winter much of the hydrogen is hidden below a layer of carbon dioxide **ice**.

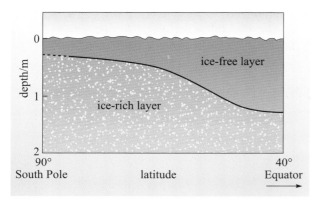

Figure 5.35 A cross-section of the near-surface regions of the martian southern hemisphere as implied by the GRS data. The ice-rich layer is approximately 0.6 m below the surface at a latitude of 60° S and extends to within 0.3 m of the surface at 75° S.

5.3 The search for life on Mars

5.3.1 Introduction

So far in this chapter you've seen evidence for past and possibly recent water flow on Mars and evidence that life on Earth can exist in a range of apparently hostile environments. These are comparatively recent scientific developments and they are two of the main reasons why the scientific exploration of Mars and the search for past or present life is a priority for both the European and United States space agencies. But there is also other intriguing evidence relevant to the question of martian life, and it comes from direct study of samples of martian material on the planet itself and from meteorites that are believed to come from Mars. However, as you'll learn in this section, detecting unequivocal evidence for life is far from easy.

5.3.2 Viking: the first search for life

By the mid-1970s, with the initial spacecraft surveys of Mars completed, NASA decided that the time was right for a major mission to Mars that would specifically address the issue of life. The answer was the Viking project, which consisted of two identical spacecraft, Vikings 1 and 2 (Section 2.3). Each consisted of an orbiter and lander, with the latter including a suite of experiments designed specifically for studying past or present life on Mars.

> **BOX 5.6 PHOTOSYNTHESIS AND CARBON FIXATION**
>
> **Photosynthesis** is the chemical processes that green plants, algae and many bacteria use to convert carbon dioxide into 'food' (i.e. complex organic molecules) and oxygen using energy absorbed from sunlight. During the first part of the process, light is absorbed by light-sensitive pigments (for example the molecule chlorophyll that gives leaves their green colour). The light energy is then used to split water into hydrogen and oxygen. The hydrogen is used to build a supply of chemical energy inside cells that is used to convert the carbon dioxide into carbohydrates. The oxygen is set free.
>
> ■ What do organisms use carbohydrates for?
>
> ❑ The simplest carbohydrates are the sugars, which are essential intermediates in the conversion of food to energy. Other carbohydrates serve as energy stores or are constituents of plant cell walls (see Section 5.1.3).

The process of photosynthesis can be summarized in a simplified chemical equation as follows:

carbon dioxide + water + light energy ⟶ carbohydrates + oxygen (released)

The carbon dioxide used by plants typically comes from the atmosphere or from carbon dioxide dissolved in water, though marine plants can use other sources of carbon.

■ How might photosynthesis affect the amount of oxygen in a planet's atmosphere?

❑ Since light energy is used to split water into hydrogen and oxygen and the oxygen is set free, photosynthesis will lead to more oxygen in a planet's atmosphere.

The conversion of carbon dioxide to carbohydrates by photosynthesis is one example of a process known as **carbon fixation** – the primary means by which plants, algae and bacteria produce the complex organic compounds that they need for growth. Many of these complex organic compounds (i.e. the organism's food) are used as energy sources, and they subsequently undergo a series of chemical reactions that breaks them down to release energy for the organism's use – a process known as **respiration**. In most plants and animals, respiration requires oxygen, and carbon dioxide is a waste product.

It is worth noting that when organisms are in environments where sunlight is unavailable alternative mechanisms must be employed to generate organic compounds. For example, in 1977 scientists in the submersible *ALVIN* were studying a mid-ocean ridge near the Galápagos Islands in the Pacific Ocean. They discovered underwater volcanoes that were populated by a range of organisms. Seawater circulating through the hot rocks of the volcanoes dissolves chemicals within the rock. These sources of mineral-rich hot seawater support communities of organisms in which life depends not on light energy, but on chemical energy.

Figure 5.36 The trench dug to sample the martian surface by the robotic arm on the Viking 1 lander. The trench is about 8 cms wide, 5 cms deep and 15 cms long.

The design of the instrument package carried on the landers was based on the assumptions that martian life, if it exists, would be carbon-based, its chemical composition would be similar to that of terrestrial life, and it would most likely metabolize simple organic compounds. It consisted of three experiments known as the Viking biology experiments designed to detect metabolic activity of potential microbial soil communities. They were:

- the pyrolytic release experiment, which tested for carbon fixation (Box 5.6).
- the gas exchange experiment, which tested for metabolic production of gaseous by-products in the presence of water and nutrients, as produced during respiration (Box 5.6).
- the labelled release experiment, which tested for metabolic activity.

Metabolism refers to all the chemical processes that occur within an organism.

Other instruments were capable of detecting organic compounds in the martian soil and could analyse the chemical composition of the martian surface soil. The Viking landers sampled the martian soil using a robotic arm (Figure 5.36).

Of the three Viking biology experiments, only the pyrolytic release experiment simulated actual martian surface conditions and did not use water. In this experiment, a small soil sample was warmed in a simulated martian atmosphere (carried from Earth), in which the carbon dioxide molecules contained some radioactive atoms. A lamp provided simulated sunlight. After five days, the atmosphere was removed and the soil sample was heated to break down any organic material (heating in the absence of oxygen is known as pyrolysis).

Labelling is the process of replacing a stable atom in a compound with a radioactive one of the same element to enable its path through a biological or mechanical system to be traced by the radiation it emits.

The resulting gases were passed through a detector to see if any organisms had ingested the radioactive carbon dioxide.

The gas exchange experiment sought to detect alterations in the composition of the gases in the test chamber as a result of biological activity. The procedure involved partially submerging a small sample of soil in a complex mixture of compounds that the investigators called 'chicken soup'. The soil was then warmed for at least 12 days in a simulated martian atmosphere containing mainly carbon dioxide. Gases that might have been emitted from organisms consuming the nutrient were then passed to a detector that was able to isolate and identify carbon dioxide, oxygen, methane, hydrogen and nitrogen.

The labelled release experiment moistened a small sample of soil with a small amount of nutrient consisting of distilled water and organic compounds. The organic compounds had been labelled with radioactive tracer atoms. After moistening, the sample was kept warm for at least 10 days. The idea was that the micro-organisms would consume the nutrient and give off gases containing the radioactive tracer, which would then be detected.

Ironically, it was the instrument designed to detect organic compounds in the martian soil rather than any of the biology experiments that arguably produced the most important result for the detection of life. It discovered no sign of any organic compound on the surface of Mars. This result came as a complete surprise as organic compounds are known to be present in space (for example, in meteorites). Proof that the instrument was working came from the fact that it was able to detect traces of the cleaning solvents that had been used to sterilize it prior to launch.

The total absence of organic material on the martian surface made the results of the biology experiments equivocal, since those experiments were designed to detect metabolism involving organic compounds. However, the results from the biology experiments are sufficiently intriguing to be worth examining.

To reduce the chance of erroneous positive results, the biology experiments not only had to detect life in a soil sample, they had to *fail* to detect it in another soil sample that had been heat-sterilized (the so-called *control* sample). The results of tests of the Viking biology experiments for both terrestrial life and samples of martian soil are shown in Table 5.4.

Table 5.4 Results of tests of the Viking biology experiments on terrestrial life and martian soil.

	Gas exchange experiment	Labelled release experiment	Pyrolytic release experiment
terrestrial life:			
sample response	oxygen or carbon dioxide emitted	labelled gas emitted	carbon detected
control response	no response	no response	no response
martian soil:			
sample response	oxygen emitted	labelled gas emitted	carbon detected
control response	oxygen emitted	no response	carbon detected

- If there was no life on Mars, now or in the past, what results would have been expected for each experiment?

☐ There should have been no response for any of the Viking biology experiments (i.e. the result that was obtained for the heat-sterilized terrestrial sample).

In fact both the gas exchange and pyrolytic release experiments produced positive results on martian soil even with the control sample. This strongly suggests that non-biological processes were influencing the responses obtained. Subsequent laboratory experiments on Earth involving the exposure to ultraviolet radiation in the presence of a martian-type atmosphere of materials thought to be similar to martian soil reproduced the results of the Viking lander experiments. It was found that iron oxide (rust) could act as a catalyst to produce the results seen by the pyrolytic release experiment and it is likely that iron oxide is the dominant form of iron in martian soil, giving the surface its red colour.

Only the labelled release experiment appears to have met the criteria for life detection (i.e. a response for the martian soil sample but not the martian soil control), but it does this rather ambiguously. When the nutrient was first injected, there was a rapid increase in the amount of labelled gas emitted. Subsequent injections of nutrient caused the amount of gas to decrease initially (which is surprising if biological processes were at work) but then to increase slowly. No response was seen in the control sample sterilized at the highest temperature. While there is still some controversy, the consensus is that the labelled release experiment results can also be explained non-biologically.

To summarize, all three Viking biology experiments gave results indicative of active chemical processes when samples of martian soil were subjected to a series of experiments designed to detect life. However, the experiments failed to detect any organic matter in the martian soil, either at the surface or from samples collected a few centimetres below the surface.

5.3.3 Meteorites from Mars

Information from martian meteorites has greatly enhanced our understanding of the chemistry and geology of Mars and has been of particular importance in interpreting the data returned by the Viking and Mars Pathfinder landers. To date (October, 2003) some 28 meteorite specimens have been identified that are believed to come from Mars, though the number is steadily increasing. One is the meteorite EET A79001 that was collected from Antarctica (Figure 5.37).

All of the martian meteorites are igneous rocks formed by the cooling of magma. However, one important fact distinguishes most martian meteorites – they have relatively young ages; EET A79001 was formed about 180 million years ago.

Figure 5.37 The meteorite EET A79001.

'EET' refers to the collection site (which, in this instance, was Elephant Moraine in the Antarctic), 'A' designates the collection trip and '79' is the year of collection (1979). The identifying number, 001, signifies that it was the first meteorite to be classified upon return of the samples to the curatorial facility.

■ The ages obtained for most of the martian meteorites suggest that they formed from magmas between about 180 million and 1300 million years ago. What are the implications of this?

❏ These meteorites were formed late in the history of the Solar System. Wherever they were formed, at least some part of the object from which they came had to have been melted 180 million years ago (or remained molten until this time).

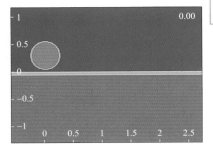

Most meteorites have ages that cluster around 4500 million years ago. They are samples of asteroids that were heated and cooled relatively early in the history of the Solar System. The only reasonable environment that retained sufficient heat to produce melting 180 million years ago is a body of planetary dimensions, so there can only be a few candidates for the source of martian meteorites, i.e. Mercury, Venus, Earth, the Moon, Mars or Io.

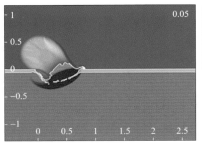

■ Can you think of a way in which the martian meteorites could be removed from the surface of the planet?

❏ Ejection caused by an impact.

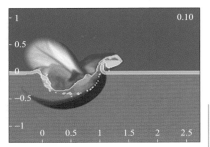

The most plausible mechanism involves removal following an impact onto a planetary-sized surface by an asteroid or comet. It is obviously not possible to do an experiment to determine the feasibility of ejecting material from Mars. Instead scientists use complex computer models to study the energy and trajectories of impacting objects. The results of one such computer model are shown in Figure 5.38. In this model, the incoming projectile, in addition to pulverizing the target rocks and producing a crater, also causes ejecta to be propelled away from the impact site. Some of these pieces of ejecta will escape from the planet's surface and some of the escaping ejecta components could be melted or shocked.

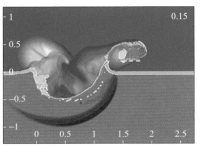

■ If fragments ejected from another body reach the Earth, what is the most likely source? Why?

❏ The Moon – simply because it is so much closer than any other solid body.

About 50 meteorites of lunar origin have now been unambiguously found on the Earth.

■ Why can we be reasonably certain that these fragments originate from the Moon?

❏ The Moon is the only body for which there are returned samples. These have been chemically analysed and compared with the fragments.

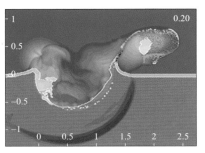

Figure 5.38 Snapshots at 0.05 second intervals showing the simulated impact of a 0.5 km-diameter body at an angle of 30° to the horizontal at 10 km s^{-1}. The vertical distance from the impact point is on the y-axis and horizontal distance on the x-axis. At 0.10, 0.15 and 0.20 seconds after impact you can see the crater opening out and ejecta expanding away from the impact direction. In this simulation, red denotes melted material from the topmost (pale green) layer in the target, evidence that glassy materials may originate from near the surface of the target. Close to the impact site, ejecta may be travelling sufficiently fast to be ejected from the planet.

Having established that some meteorites found on Earth have originated from another body in the Solar System, we can now turn again to EET A79001. There is some very specific evidence that quite unambiguously points to Mars as the source of EET A79001. You've already seen that the martian meteorites are the result of igneous activity, analogous in many ways to rocks formed at or near the Earth's surface. However, martian meteorites were also subjected to an impact event. Intuitively therefore, it might be expected that the meteorites would show some evidence of this process. Indeed, some martian meteorites record the effects of shock. During the impact event that is thought to have ejected EET A79001, localized melting occurred within the sample. The melt cooled very rapidly to form a glass containing trapped atmospheric gases. Analyses of these trapped gases showed them to be chemically distinctive. The abundances of the gases contained within the shock-produced glass from EET A79001 are plotted in Figure 5.39 against the abundances of gases in the Martian atmosphere as determined by the Viking and Pathfinder missions. The line represents points whose abundance values are the same on both the vertical and horizontal axes.

EET A79001 was ejected from Mars in an impact event about 600 000 years ago.

■ What can you conclude from Figure 5.35 about the abundances of gases in EET A79001 and the Martian atmosphere?

❑ They look the same, i.e. the data points plot on the line where abundance values would be the same on Mars and EET A79001.

The observation that gases have the same abundances in EET A79001 as the martian atmosphere strongly suggests that EET A79001 contains trapped martian atmosphere and therefore comes from Mars.

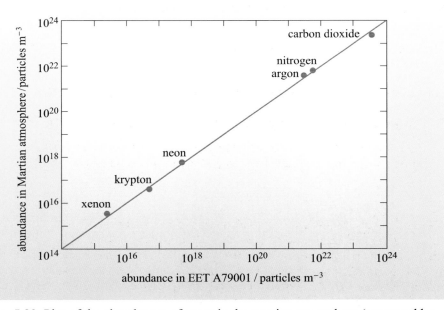

Figure 5.39 Plot of the abundances of gases in the martian atmosphere (measured by spacecraft landers) versus the abundances in the glass from EET A79001.

ALH 84001: evidence of life in a Martian meteorite?

You now know why scientists believe that we have samples of Mars available for study in the laboratory in the form of martian meteorites. In this final part of our look at evidence for past or present life on Mars we'll examine the most famous of all martian meteorites and its significance for the question of life on Mars. Until 7 August 1996, the name ALH 84001 was probably known to only a relatively small number of planetary scientists worldwide.

But on that day, everything changed. A press conference was held by a group of scientists led by David McKay, Everett Gibson and Kathy Thomas–Keptra of NASA's Johnson Space Center to announce that certain characteristics in this meteorite, known to have come from Mars, were most likely interpreted as being the relics of ancient martian microbial life. The effect was dramatic. Newspapers, radio and TV around the world rushed to declare 'Life found on Mars!' (or its equivalent in many languages) and with their own particular stress. It seemed that humanity's desire to find that we are not alone (or at least might not have been alone at some time in the past) had at last been answered. But had it? Well not everyone in the scientific community thinks so. In fact, it is true to say that the majority of informed opinion does not agree with the conclusions of the authors of the scientific paper that presented the full analysis of their case.

Why did a 1.9 kg meteorite (Figure 5.40a) cause such a sensation? The *recent* history of ALH 84001 started when it was discovered in 1984 in the Allan Hills region of Antarctica. However, the martian origin of ALH 84001 was not recognized until 1993. But what of the history of ALH 84001 before it was found in Antarctica in 1984? Here, we shall quote from a 1997 article by several of the team who undertook the original work:

> 'The meteorite timeline begins with the crystallization of the rock on the surface of Mars, during the first 1 percent of the planet's history. Less than a billion years later the rock was shocked and fractured by meteoritic collisions. Some time after these impacts, a water-rich fluid flowed through the fractures, and tiny globules of carbonate minerals formed in them. At the same time, molecular by-products, such as hydrocarbons, of the decay of living organisms were deposited in or near the globules by that fluid. Impacts on the surface of Mars continued to shock the rock, fracturing the globules, before a powerful collision ejected the rocks into space. After falling to Earth, the meteorite lay in the Antarctic for millennia before it was found and its momentous history revealed.'
>
> Gibson, E.K. *et al.* 'The Case for Relic Life on Mars', *Scientific American*, (December 1997) pp.58–65.

ALH 84001 was ejected from Mars around 16 million years ago and landed on Earth about 13 000 years ago.

- How old is the ALH 84001 meteorite?

- According to the quotation above, ALH 84001 formed 'during the first 1 percent of the planet's history'. Since Mars, like all the planets, formed around 4500 million years ago, ALH 84001 formed around 45 million years later. This means that it has an age of nearly 4500 million years. This is in contrast to most martian meteorites which have an age in the range 180 million to 1300 million years.

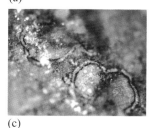

Figure 5.40 (a) ALH 84001 shown in its entirety. (The meteorite is 20 cm across.) (b) A cut through the rock, showing a cross section. Just right of centre is a vertical crack through which fluid flowed and deposited globules of carbonate minerals. (c) A fragment, 2 mm in length, containing several of the carbonate globules, which are about 0.2 mm in size.

But how did the scientists reveal this 'momentous history', namely that ALH 84001 contained evidence for the existence of living organisms early in the history of Mars? Their conclusion was based on five separate strands of evidence, each of which on its own was not compelling, but, taken together, according to the authors, were highly convincing. Almost all of their evidence comes from carbonate globules that are found on the surface of a fracture through which fluid flowed and deposited these globules (Figure 5.40b, c).

The five lines of evidence, as presented by David McKay and his colleagues can be summarized as follows:

1   The carbonate is in the form of 'globules' comparable to aggregates of similar crystals known to be produced by bacteria on Earth.

2   Perhaps the most visually compelling evidence are objects that seem to be the fossilized remains of microbes themselves. Extremely small carbonate structures in the globules resemble fossil spherical or rod-shaped bacteria (Figure 5.41a). The segmented object shown in Figure 5.41a is only 3.8×10^{-7} m in length. This can be compared with terrestrial samples such as that shown in Figure 5.41b. The approximately vertical feature just to the right of centre is believed to be a minute fossil the same size as the carbonate structures in ALH 84001. It was found 400 m below the Earth's surface in basalt.

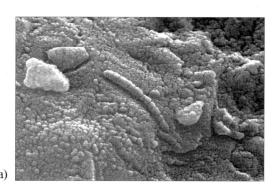

Figure 5.41 (a) The segmented object (3.8×10^{-7} m long) was found in the Martian meteorite ALH 84001 and has been interpreted as a fossil microbe. It is claimed to resemble fossilized bacteria found on Earth. For example, the near-vertical object in (b), which is of the same length as the object in (a) was found at a depth of 400 m below the surface of the Earth.

3   Inside the carbonate globules, they found fine-grained particles of the minerals magnetite (a form of iron oxide) and pyrite (iron sulfide). Using sophisticated analysis techniques the team found that the size, purity, shapes and structure of all the magnetites were typical of magnetites produced by bacteria on Earth. Such particles on Earth are known as magnetofossils. Intriguingly, some of the magnetites in ALH 84001 are arranged in chains, similar to pearls in a necklace. Free-swimming terrestrial bacteria often produce magnetite in just this pattern, because as they biologically process iron and oxygen from water, they produce crystals that align themselves with the Earth's magnetic field.

4   Another strand of evidence is that the carbonate, iron sulfide and iron oxide minerals occur together, yet they would not normally be stable under the same physical conditions. This suggests formation as a result of living processes.

5   ALH 84001 contains organic matter, including complex hydrocarbons, that could have been produced by living organisms. Organic molecules have been found in other meteorites, which are known to have come from asteroids, so why should this fact be significant? McKay *et al.* argue that the type and relative abundance of the specific organic molecules are suggestive of life processes. The organic matter is also found in carbonate-rich regions, including the globules, and contains some molecules that could be the decay products of microbes.

■ What term would we use to describe organic compounds and mineralogical signatures of past life?

❏ Biomarkers (see Section 5.1.5).

All these features occur together in the carbonate globules and the scientists argued that they were indigenous to the meteorite and greater than 3900 million years old. They therefore concluded that there was life on Mars at that time. There is however much scepticism in the scientific community as to the validity of this conclusion, and much research on this subject has been instigated since 1996. Amongst the objections to these conclusions are:

- It is always difficult to find compelling evidence of a microbial origin for simple structures of the type seen in ALH 84001, which might equally plausibly be chemical and mineralogical artefacts. The images of the possible microfossils are articles of faith – if you want to see bacteria then that's what you see, but most scientists see crystal edges and other crystal features.

- Convincing fossils of this age are extremely rare on Earth, and very hard to find even in systematic searches of well-exposed regions of well-preserved rocks. So it is very surprising that with a sample of little more than 20 rocks from Mars, one were to contain fossils.

- The original research compared the carbonate microstructures to fossil bacteria but made no attempt to make comparisons with non-biologically produced mineral structures that could be mistaken for fossil bacteria.

- More recent studies suggest that the carbonates and other materials in the veins may have been subjected to temperatures of 200 °C to 500 °C at the moment of the impact that ejected ALH 84001 into space. If the fractures and the carbonates that fill them formed this way, then the structures and compounds within them are extremely unlikely to be fossils.

It now seems that of the five original lines of evidence, four can be explained without the necessity for a biological origin. However, one still remains controversial: the origin of the magnetite crystals present in the carbonate globules. Since 1996, the original team has put forward arguments to counter the objections described above. The present consensus of the scientific community is that while the evidence for life is intriguing and demands further study, it is not compelling. It is probable that the argument will only be finally resolved by results from future missions to Mars. Perhaps the most important legacy of the debate over ALH 84001 is the impetus it has given to studies of life on Mars.

5.4 Summary of Chapter 5

- A working definition of life is that it is a self-sustaining chemical system capable of undergoing Darwinian evolution. The discovery of the structure of DNA provided the basis for understanding one of these key characteristics: the mechanism that enables biological molecules to replicate themselves.

- Carbon is important to life as it can form chemical bonds with many other atoms, allowing a great deal of chemical versatility; it can form compounds that readily dissolve in water.

- A habitable zone is defined as encompassing the range of distances from a star for which liquid water can exist on a planetary surface. Mars, with a mean distance from the Sun of 1.52 AU, falls within our Sun's habitable zone, which extends from 0.95 AU to 1.67 AU.

- Water molecules are the major component of living tissues, generally accounting for 70% of their mass. The other major molecules in a living system are large organic molecules: lipids, carbohydrates, proteins and nucleic acids.

- The continued discovery of new extreme environments and the organisms that inhabit them has made more plausible the search for life on Mars. Organisms that are capable of different degrees of adaptability to the extreme range of living conditions available on Earth are known as extremophiles.

- The northern polar cap of Mars is believed to comprise carbon dioxide overlying a residual cap of water-ice about 600 km across, which is exposed in summer. In contrast, the south polar cap appears to be composed predominantly of carbon dioxide.

- The surface of Mars is rather cold. At low latitudes, a typical daily temperature range is from $-100\,°C$ to $+17\,°C$ with a mean of about $-60\,°C$. In winter, the temperatures at the pole can fall to $-125\,°C$.

- The Viking biology experiments gave results indicative of active chemical processes when samples of martian soil were subjected to a series of experiments designed to detect life. However, the experiments failed to detect any organic matter in the martian soil, either at the surface or from samples collected a few centimetres below the surface.

- Gases trapped in some meteorites have the same abundances as the martian atmosphere strongly suggesting that those meteorites come from Mars. Evidence from the martian meteorite ALH 84001 has been interpreted by some scientists as indications of past life on Mars.

5.5 Questions

QUESTION 5.4

Which *one* of the following statements might be considered a scientific definition of life?

A   A chemical system that can make copies of itself.

B   An organic substance that evolves over time to a more complex organic substance.

C   A chemical system able to replicate itself and to make mistakes in that replication that gradually changes and increases the efficiency of the system.

D   A chemical system able to interact with its environment.

E   A chemical system that absorbs energy and becomes more complex over time.

QUESTION 5.5

Which *two* of the following chemical elements are relatively abundant in the Universe and living organisms?

A   helium

B   hydrogen

C   phosphorous

D   silicon

E   nitrogen

F   magnesium

QUESTION 5.6

Compare the types of adaptations that thermophilic and psychrophilic organisms have developed in order to exist in temperatures above 60 °C and below 0 °C.

QUESTION 5.7

Which *two* of the following statements correctly *contrast* either the composition or structure of the atmospheres of Earth and Mars?

A   The martian and terrestrial thermospheres extend to similar altitudes above the surfaces of the two planets.

B   Carbon dioxide is a major constituent of the martian atmosphere but not the Earth's.

C   Atmospheric pressure on Mars is around 1% of that on Earth.

D   There is a significant difference in atmospheric temperature at a height of 10 km above the surfaces of Earth and Mars.

E   The martian stratosphere occurs at a higher altitude than the Earth's.

QUESTION 5.8

Briefly explain why water is important in the fixation of carbon during the process of photosynthesis.

QUESTION 5.9

Briefly describe how the gamma-ray spectrometer on Mars Odyssey has been used to infer the presence and distribution of water on Mars.

QUESTION 5.10

The Viking biology experiments employed a control sample against which results from martian soil samples were compared. If this control sample had not been used, how might the results from the martian samples have been interpreted?

AFTERWORD

In reading through this course text you have examined the state of our scientific understanding of Mars as of late September 2003. Even while this course was being written, our knowledge of Mars has been changing, at times on an almost daily basis as new results are returned from the Mars Odyssey and Global Surveyor spacecraft currently in Mars orbit. For example, Mars Odyssey acquired the map you examined in Figure 5.34 showing the distribution of hydrogen after collecting more than a year's worth of data. The map was published at the end of July 2003. Such rapid developments are common in planetary science during the active science stages of space missions. However, work on the missions themselves starts years, and at times decades, before any scientific return. The investment in time and resources is enormous and not without risk – the exploration of Mars has proven challenging, as a quick glance of the unsuccessful missions in Table 2.1 will show.

Not surprisingly therefore, June and July of 2003 were both busy and anxious times for mission scientists and engineers in Europe and the United States. On 2 June ESA's Mars Express mission carrying the Beagle 2 lander was launched from Baikonur in Kazakhstan, followed by the launches of NASA's two Mars Exploration Rovers: Spirit on 10 June and Opportunity on 7 July from Cape Canaveral in Florida.

In this final section, we'll take a brief look at some of the scientific objectives of the Mars Express and Mars Exploration Rover missions and some of the plans for possible future missions to Mars.

The Mars Express mission

Mars Express represents Europe's first venture to the red planet. It's an ambitious mission involving both an orbiter and a planetary lander. The orbiter carries an array of instruments to examine the surface and atmosphere of Mars. What is distinctive about this package is the MARSIS instrument (Mars Advanced Radar for Subsurface and Ionospheric Sounding), a ground-penetrating radar. As the name implies, this has the ability to study details below a surface using radio waves that are capable of penetrating the surface. Such an instrument has been deployed only once successfully in space, on one of the Apollo lunar orbiting missions. This is the first time that such an instrument will be deployed at Mars and the first time specifically to search for water.

MARSIS beams low-frequency radio waves towards the planet from an aerial, 40 m in length, due to be unfurled after Mars Express goes into orbit. The radio waves will be reflected from any surface they encounter. For most, this will be the surface of Mars. But some penetrate the surface to encounter further layers of different materials so, for instance, it will be possible to measure the thickness of sand dunes. If there is a layer containing liquid water, it should generate a specific radar signal. Consequently it is anticipated that MARSIS should be able to pick out layers of rock interspersed with ice, or determine whether there are layers of sediment sitting on top of other material in areas that may have been the sites of ancient lakes or oceans.

The Beagle 2 lander, with a landed mass of less than 30 kg represents the most ambitious ratio of science payload to total spacecraft mass ever attempted. Nearly one-third of the entire probe mass is devoted to various types of analysis or is used to manipulate and collect samples for study on the surface of Mars. The remaining

Beagle 2's landing site is in Isidis Planitia at approximately 10° N, 90° E.

mass of the lander is used for the power system (batteries), telemetry to relay the data, thermal protection to ensure the delicate instruments don't get too hot or cold and an overall structure.

The lander carries five instruments designed to detect, directly or indirectly, signs of present or past life on Mars and to study the martian surface environment. One of these is the Gas Analysis Package (GAP) which is designed to study samples of the martian atmosphere, as well as soil and rock. There are two devices for collecting samples for GAP (and some of the other instruments). These are a 'mole' which is able to collect material from on and below the surface and a corer–grinder, which is able to drill into rocks and collect fresh material from inside rocks. Once this material is delivered, GAP will concentrate on examining the samples for possible biomarkers.

Mars Exploration Rovers

The first of NASA's two Mars Exploration Rovers (MERs) are scheduled to land on Mars in early January 2004 after a seven-month flight, the second will follow a few weeks later. The landers have a design that is dramatically different from previous Mars landers; each MER spacecraft carries a single long-range rover with a mass of nearly 180 kg and a range of up to 100 metres per martian day. After landing, each rover undertakes a reconnaissance of its landing site using both visible and infra-red images prior to scientists back on Earth choosing potential rock and soil targets for further investigation. When a rover reaches a target, its multi-jointed arm will deploy and the target will be examined with a microscope and two spectrometers. The 'RAT' (Rock Abrasion Tool) will be used to expose fresh rock surfaces for study.

The main goal of the mission is to determine the history of climate and water at sites on Mars where conditions may once have been favorable to life. The proposed landing sites at Gusev Crater (15° S, 175° E) and Meridiani Planum (0° N, 358° E) have been chosen using data from Mars Global Surveyor spacecraft and other missions since they offer evidence that liquid water was once present. Surface operations are planned to last for at least 90 martian days, extending into the early summer of 2004.

Future missions

A number of missions have been proposed for the further exploration of Mars over the next two decades. You should note, however, that some of these missions are still little more than ideas while others are in the early stages of planning. NASA's next planned mission to Mars after the Mars Exploration Rovers is the Mars Reconnaissance Orbiter due for launch in 2005. This mission will focus on analyzing the surface at even greater resolution to explore the tantalizing hints of water detected in images from the Mars Global Surveyor spacecraft, and to bridge the gap between surface observations and measurements from orbit. Other NASA missions proposed, together with their possible launch dates include:

- Phoenix (2007), a polar lander designed to replace the ill-fated Mars Polar lander lost in 1999.
- Mars Science Laboratory (2009), a proposal for a long-duration, long-range mission to explore the martian surface ahead of a possible sample return mission.

- Mars Sample Return Lander (2011–2014), a mission proposal to return samples of the martian surface to Earth where they can be studied using the full range of laboratory techniques available.

ESA's future strategy for space exploration falls within its Aurora programme, the goals of which are to define, and later implement, a strategy for the robotic and human exploration of the Solar System. Aurora will define the missions, technologies and ground-based activities necessary to implement the strategy. ESA's next proposed mission to Mars, ExoMars (2009), includes a rover that will search for signs of life, past or present, determine the distribution of water on Mars, measure the chemical composition of the surface rocks and identify potential surface hazards to future human missions. Ultimately, Aurora aims to prepare Europe for a prominent role in the international endeavour to send the first humans to Mars (which is likely to be at least 30–40 years away).

Space missions are expensive and risky – exactly how the future exploration of Mars will pan out is dependent upon all kinds of financial, political and technical issues. However, our understanding of Mars may be influenced in ways that do not involve missions at all. Meteorites from Mars arrive on Earth for free. A quantum leap in the development of ideas about Mars may take place tomorrow, from a lump of rock which lands in your street!

ANSWERS AND COMMENTS

QUESTION 1.1

The *incorrect* statement, which is therefore the option you should have selected, is C. The atmosphere of Jupiter is composed of 90% hydrogen and 10% helium. This information is given in Section 1.2.1.

QUESTION 1.2

This question is designed to familiarize you with many of the feature names you will meet in your study of Mars. (a) Isidis Planitia is a low-lying plain; (b) Syrtis Major Planum is a high plateau or high-lying plain; (c) Gordii Dorsum is a ridge-like feature; (d) Orcus Patera is an irregular-shaped crater.

QUESTION 1.3

This question is asking you to summarize some of the information that you have met in your reading of Chapter 1. A useful approach to this type of question is to go through the text and highlight relevant points.

(a) The main features of the atmospheres.

Mercury has a very tenuous atmosphere, it is 1×10^{-15} times less dense than Earth's. Venus has a thick atmosphere composed of 97% (by volume) carbon dioxide that causes a significant greenhouse effect. Earth has an atmosphere composed of 78% (by volume) nitrogen and 21% (by volume) oxygen plus a small amount of other gases. Mars has a rather tenuous atmosphere, roughly 0.006 times that of Earth's atmosphere composed of 95% (by volume) carbon dioxide.

(b) Surface temperatures.

Mercury experiences extremes of temperatures (from 470 °C to –120 °C) due to its close proximity to the Sun. The important point about the surface temperatures of Venus, Earth and Mars is that they are all influenced by the extent of the atmospheric greenhouse effect on each planet. Venus's thick atmosphere of carbon dioxide has resulted in a substantial greenhouse effect that raises surface temperatures to around 400 °C. Earth also has an average surface temperature (15 °C) that is 33 °C higher than an atmosphereless Earth would have because of the greenhouse effect. Mars, despite its carbon dioxide atmosphere, is further from the Sun so its average surface temperature (–50 °C) is only slightly affected by a greenhouse effect.

(c) Surface topography.

The surface of Mercury is dominated by impact craters, the high number of which suggests that its surface has not been geologically active. The Magellan mission to Venus used radar to penetrate its thick atmosphere and found evidence of a geologically active surface dominated by volcanoes (some 85% of its surface has been affected by volcanic processes). Earth's surface is obviously dominated by liquid water. The lack of impact craters tells us this is a planet with a dynamic surface, including large volcanoes such as Mauna Kea in Hawaii. Mars shows some extremes of topography with large canyons such as the Valles Marineris and giant volcanoes such as Olympus Mons, evidence that Mars has also been a geologically active world.

QUESTION 1.4

The two *incorrect* statements, which are therefore the options you should have selected, are D and E. Valles Marineris is considerably larger than the Earth's Grand Canyon. The first spacecraft to visit Mars was Mariner 4 (Mariner 2 visited Venus).

QUESTION 1.5

You could describe Giovanni Schiaparelli's meticulous telescope observations of Mars documenting channels and grooves on its surface. In Italian, the word 'canali' was used, which led to Schiaparelli's observations being interpreted incorrectly as canals, with the obvious implications for martian life that such a conclusion entailed (i.e. that intelligent life had built the canals).

QUESTION 2.1

In either direction, across or down the image, the resolution is given by the image size divided by the number of pixels.

So, top to bottom:

$$\text{resolution} = \frac{4.5\,\text{km}}{512} = 0.0088\,\text{km} = 8.8\,\text{m}$$

And, side to side:

$$\text{resolution} = \frac{12.7\,\text{km}}{1024} = 0.124\,\text{km} = 12.4\,\text{m}$$

(a) (i) The resolution in both directions is much smaller (and therefore better) than 500 m. So impact craters of this size would be distinguished (resolved).

(ii) Conversely, in this case, 1 m is below the figure for the resolution in both directions. Therefore, 1 metre-scale boulders would not be resolved.

(b) In this case, the area covered by the image would be larger and the same resolution would cover a greater area, so the detail would be less.

QUESTION 2.2

The two *incorrect* statements, which are therefore the options you should have selected, are A and E. Mars Odyssey's gamma-ray spectrometer detects gamma-rays, which have a wavelength of around 10^{-12} m (the correct statement given in D). Mariner 4 sent its data back to Earth using radio waves, which have a wavelength of between 10^0 m and 10^3 m not 10^{-3} m.

B is correct since 10^{-5} m is a typical wavelength for infra-red radiation (detectable by Mars Odyssey's THEMIS instrument). The wavelength given in C (6×10^{-7} m) is in the visible region of the electromagnetic spectrum and would be detectable by Mars Global Surveyor's optical camera.

ANSWERS AND COMMENTS

QUESTION 3.1

As the planet has equal quantities of iron and silicate rock, we can determine its mean density by adding the densities of iron and silicate rock and dividing by two. Thus:

$$\text{mean density} = \frac{7.9 \times 10^3 \text{ kg m}^{-3} + 2.7 \times 10^3 \text{ kg m}^{-3}}{2} = 5.3 \times 10^3 \text{ kg m}^{-3}$$

QUESTION 3.2

From Table 3.1: mass of Mars = 0.642×10^{24} kg

Number of planetary embryos required to assemble Mars =

$$\frac{\text{mass of Mars}}{\text{mass of planetary embryos}} = \frac{0.642 \times 10^{24} \text{ kg}}{5 \times 10^{22} \text{ kg}} = 12.8 \quad (\text{i.e. } 12-13 \text{ planetary embryos})$$

From Table 3.1: mass of Earth = 5.97×10^{24} kg

Number of planetary embryos required to assemble Earth =

$$\frac{\text{mass of Earth}}{\text{mass of planetary embryos}} = \frac{5.97 \times 10^{24} \text{ kg}}{5 \times 10^{22} \text{ kg}} = 119.4 \quad (\text{i.e. } 119-120 \text{ planetary embryos})$$

QUESTION 3.3

The density of basalt (3×10^3 kg m^{-3}) is lower than the mean density of Mars (3.93×10^3 kg m^{-3}). This implies that the density at depth in Mars must be greater in order for the mean density to be the calculated value.

QUESTION 3.4

Your summary could include the following points. The Solar Nebula consisted of a cloud of dust and gas that contracted due to mutual gravitational attraction, eventually forming a spinning protoplanetary disc, with the young Sun at its centre. Within the disc, material began to accrete. Initially, dust particles collided and stuck together gradually forming larger and larger particles from a few millimetres to a few metres in size. Continued accretion led to the growth of planetesimals 0.1–10 km across, some of which grew at a faster rate, sweeping up smaller planetesimals, to form planetary embryos a few thousand kilometres across.

Collisions between planetary embryos, and their associated impact melting, allowed differentiation to occur with denser material segregating towards the centre of the embryos. In total, Mars probably took around 100 million years to accrete.

QUESTION 3.5

The *correct* statement is C. Planetary differentiation is the process whereby the interior of a planetary body segregates into layers of different compositions (see Box 3.3).

QUESTION 3.6

From Table 3.1, the *correct* answer is B. Earth has the highest density of the terrestrial planets and Mars the lowest.

QUESTION 3.7

The two *correct* statements are C and E. Evidence from Mars Global Surveyor suggests that both planets have a partially liquid core. As both planets formed by the same processes of planetary accretion and differentiation it is likely that they both have mantles composed of the rock peridotite.

QUESTION 3.8

The *correct* statement is B. Comparison with sand dunes on Earth would suggest that martian sand dunes are most likely the result of wind-blown fine sediment and are therefore sedimentary rocks. It is not possible to determine from the image whether the dunes are still loosely bound together (and therefore not actually rocks) or whether they were formed a long time ago and are now solid rock.

QUESTION 4.1

The two *correct* statements are A and B. Both these processes rely on convection to remove the heat. As space is a vacuum, then the loss of heat from the spacecraft cannot be by convection. Similarly, an unmanned spacecraft will not warm by convection – though convection can occur in the atmosphere inside a manned spacecraft. The transfer of heat through the walls of a house is by conduction.

QUESTION 4.2

You might mention that the strongest evidence in favour of some form of plate tectonics having operated at some time in Mars' history comes from the preserved magnetic stripes in regions of its crust. On Earth such stripes are produced by the movement of crustal material away from mid-ocean ridges and a similar process could have produced the stripes on Mars. Evidence that strongly suggests that plate tectonics is *not* now operating on Mars comes from the thickness of the martian lithosphere and the shear size of some of its volcanoes. The size of the volcanoes suggests that they have been built up over a stationary hot spot in the martian mantle over a long time period, in contrast to similar volcanoes on Earth where plate tectonics moves oceanic crust over hot spots.

QUESTION 4.3

The *correct* statement is D, Mars has cooled more quickly than the Earth. The major control on volcanic activity is the amount of primordial and ongoing heat sources within the planet. Tidal heating of the Earth by the Moon will generate a very small amount of extra heat.

QUESTION 4.4

Because the pit is so small, the meteorite that formed it was clearly a small one and its speed of impact must have been severely limited by the Earth's atmosphere. It therefore more nearly resembles a pebble thrown into mud rather than an explosive high-speed impact. The asymmetrical distribution of ejecta suggests that the meteorite was travelling from right to left, and it probably struck the road at an oblique angle.

QUESTION 4.5

The *correct* statement is B, the diameter of the transient cavity will be larger than the projectile (see Figure 4.30).

QUESTION 4.6

Features you might mention that distinguish impact craters from volcanic craters are:
- impact craters are surrounded by a blanket of ejecta
- the impact crater floor is lower than ground-level beyond the crater
- the inner wall of the impact crater has slumped to produce a series of terraces
- complex impact craters have a central uplift.

QUESTION 4.7

Characteristics you might mention include:
- The enormous size of martian volcanoes compared to terrestrial ones. You could also explain that this is a consequence of the thicker lithosphere on Mars and the lack of plate tectonics, which keeps martian volcanoes over their hot spots allowing them to grow to bigger sizes.
- Martian volcanoes are predominantly shield volcanoes, characterized by gentle slopes and broad calderas, showing evidence of effusive volcanism rather than explosive volcanism.
- The largest volcanic province on Mars, Tharsis, is characterized by a large 'bulge' in the planet. Martian volcanoes are not associated with plate boundaries.

QUESTION 4.8

The *incorrect* statement, and therefore the option you should have chosen, is B. Radiogenic heating is an ongoing source of internal heat in planets, not a primordial heat source (see Box 4.1).

QUESTION 4.9

The two *correct* features would be C and E (from Table 1.1). Both Paterae and Tholi are recognized as volcanic landforms on Mars (see Section 4.2.3).

QUESTION 4.10

Your summary might include the following points, listed in order.

There are three distinct stages to the impact process. The initial stage (contact and compression) commences when the projectile makes contact with the target's surface at a speed of a several kilometres per second. The kinetic energy of the projectile is largely transferred to the target. The next stage (excavation) commences with roughly hemispherical shock waves surrounding the projectile, which weaken and shatter the target and begin excavating material outwards to produce a transient cavity. Once the shock waves have moved beyond the rim of the developing crater they take no further part in the impact process and we enter the final stage (modification). The extent of modification depends on the size of the crater, small craters preserve the shape of the transient cavity with minor modification, large craters cannot sustain the transient cavity so the rocks on the edge of the crater collapse inwards.

QUESTION 4.11

The *correct* answer is E, all of the processes *could* produce layered deposits. However, scientists generally believe that the most likely process for many of the layered deposits is the action of water, i.e. fluvial processes. Confirmation, however, will have to await direct geological observations on the surface of Mars.

QUESTION 4.12

Your completed table should resemble the one below.

Process	Earth	Mars
impacts	1	2
volcanism	2	3
erosion (fluvial and aeolian)	3	1–2

You may have chosen to separate fluvial and aeolian processes. Both would rank as high on Earth; aeolian processes would perhaps rank as medium on Mars and fluvial processes would perhaps rank as low to medium.

QUESTION 5.1

Some things you might consider for your definition are:

Living things need energy.

Living things grow and develop.

Living things respond to their surroundings.

Living things reproduce.

You might also have considered life's complexity. However, complexity by itself is not enough; much of the Universe is complex but non-living – after all, a dead mouse is, for a little while at least, almost as complex as a living one.

QUESTION 5.2

The *correct* answer is C, a psychrophile. That type of extremophile would be more likely to survive the low temperatures of an Antarctic environment.

QUESTION 5.3

The extreme range between day and night temperatures on Mars is the result of heat loss from the side of the planet that is not facing the Sun. Because of Mars' thin atmosphere the surface heats up rapidly during the day because the overlying atmosphere absorbs very little solar radiation. However, at night heat from the surface readily escapes from the planet, again because the atmosphere is so thin.

QUESTION 5.4

The *correct* statement is C, a chemical system that can *both* replicate and evolve. The other options are partial definitions of some of the properties of life.

QUESTION 5.5

The two *correct* answers are B and E, hydrogen and nitrogen (see Table 5.1).

QUESTION 5.6

There are some similarities in the style of adaptations of psychrophilic and thermophilic organisms that enable them to respond to extremes of temperature, although the adaptations themselves are distinct. Both have adapted their cell membranes to enable them to function at high and low temperatures (Table 5.3). The proteins and nucleic acids of thermophiles have adapted so that they are stable and do not breakdown at high temperatures, while the enzymes of psychrophiles function optimally at low temperatures.

QUESTION 5.7

The two *correct* statements are B and C. A is incorrect because the thermosphere on Mars extends to a greater height than on Earth (Figure 5.22). D is incorrect because there is no significant difference in temperature at a height of 10 km in either atmosphere (Figure 5.22). E is incorrect because the stratosphere is unique to Earth.

QUESTION 5.8

Water is important in photosynthesis because it provides hydrogen that is used to build up a supply of chemical energy that is used to convert carbon dioxide into carbohydrates.

QUESTION 5.9

Because water is a molecule consisting of hydrogen and oxygen, Mars Odyssey used its gamma-ray spectrometer to map the distribution of the element hydrogen across the surface of Mars. Hydrogen was chosen rather than oxygen since most minerals contain oxygen. The gamma-ray spectrometer can detect the gamma-rays produced when cosmic rays interact with the nuclei of different chemical elements.

QUESTION 5.10

The results from the gas exchange, labelled release and pyrolytic release experiments from the martian samples indicated, respectively, that oxygen was emitted, labelled gas was emitted and that carbon was detected. These are superficially the same results that would be expected from terrestrial life samples (see Table 5.4). So without the control samples, which would have shown similar results, at least for the gas exchange and pyrolytic release experiments, the Viking biology experiment results might have been interpreted as indicative of the existence of life.

GLOSSARY

accretion The growth of bodies during the formation of the Solar System, as a result of collisions that are not sufficiently energetic to fragment and disperse the colliding bodies. The term is usually employed to describe the growth of bodies from planetesimal-size upwards, and is used irrespective of whether the colliding bodies are roughly equal in mass or whether one is much smaller.

acidophiles An extremophile that thrives in conditions of acidity, typically between 0.7 and 4 on the pH scale.

advection The transfer of heat by the movement of magma through a planetary body.

aeolian Applied to the processes of erosion, transport and deposition of material due to the action of the wind at or near a planet's surface.

alkaliphiles An extremophile that thrives in conditions of alkalinity, typically between 8 and 12.5 on the pH scale.

asteroid belt Region between the orbits of Mars and Jupiter, from about 2.0–3.3 AU, where the majority of asteroids are found. Also referred to as the asteroid *main belt*.

asthenosphere A weak convecting zone of a planetary body, underlying the lithosphere.

astronomical unit The astronomical unit is a convenient measure of distance within the Solar System. It is the mean distance between the (centres of) the Sun and the Earth, and is about 149.6 million kilometres.

atmospheric structure The broadly horizontal layering of a planet's atmosphere.

basalt A dark, fine-grained extrusive igneous rock.

breccias Rock composed of sharp-angled fragments embedded in a fine-grained matrix. Brecciated rocks are commonly associated with the impact cratering process.

brecciated See breccias.

calderas Large volcanic craters, more than 1 km in diameter, formed by subsidence following the eruption of large volumes of lava or pyroclastic rocks. Major basaltic volcanoes such as Mauna Loa on the Earth and Olympus Mons on Mars are commonly crowned by calderas.

carbohydrates Collective term for sugars and polysaccharides, which are chain-like molecules made up of carbon, hydrogen and oxygen.

carbon fixation The process by which carbon dioxide from the atmosphere is assimilated into organic compounds in living organisms.

carbonates Minerals containing carbon and oxygen and usually metallic elements such as calcium or magnesium.

core The central part of a planetary body.

crust A compositionally distinct outer layer that can be recognized on top of the mantle on evolved planetary bodies such as Mars. It forms the outermost part of the lithosphere.

cryovolcanism Volcanic processes taking place at low temperatures involving icy magmas with melting points less than 0 °C, rather than silicates, which melt at temperatures exceeding 700 °C. Several icy satellites of the outer planets display evidence of cryovolcanism.

cytosol An internal aqueous component of a cell.

Darwinian evolution The theory that evolution occurs by the natural selection of individuals with characteristics that enable them to reproduce successfully and pass on their traits to their offspring.

differentiation The process whereby a planetary body (large planetesimal, planetary embryo, planet or satellite) evolves into compositionally distinct layers. Dense material sinks to form the core, and less-dense material rises to form the mantle. Sometimes, a chemically distinct crust forms on top of the mantle.

DNA (deoxyribonucleic acid) Polymer found in all cells, and in some viruses, which is responsible for forming the genetic code. It consists of two long chains of alternating deoxyribose sugar molecules and phosphate groups linked by nitrogenous bases.

double helix The configuration of the two strands of the DNA molecule: two parallel helices wound about a common axis.

ecliptic plane The plane of the Earth's orbit around the Sun. The orbits of the planets, apart from Pluto, lie very near the ecliptic plane.

effusive volcanism A volcanic eruption in which the dominant products are lava flows rather than pyroclastic rocks.

electromagnetic radiation A flow of energy consisting of radiation from all or part of the electromagnetic spectrum.

electromagnetic spectrum A collective term used to describe the various wavelength ranges of electromagnetic radiation. In order of increasing wavelength, these ranges are gamma (γ) rays, X-rays, ultraviolet radiation, visible light, infrared radiation, microwaves and radio waves.

endoliths An extremophile that lives on or inside rock or in the pores between mineral grains.

enzymes Catalysts found in living organisms that allow complex biochemical reactions to occur. Most enzymes are proteins.

erosion The removal of part of the surface of a planet by wind, water, gravity or ice. These agents can only transport matter if weathering has first broken up the material.

eruption columns Column of hot gas, ash and dust which result from an explosive volcanic eruption.

explosive volcanism A volcanic eruption in which the dominant products are usually fragmented solid materials (pyroclastic rocks) rather than flows of molten lava.

extremophiles Organisms (normally micro-organisms) that can grow under extreme conditions of, for example, heat, cold, high pressure, high salt concentrations, or high or low pH.

extrusive Rocks that have been extruded at a planet's surface as lava or other volcanic deposits.

fire fountains Sustained sprays of incandescent basalt lava propelled by escaping gas. Fire fountains may reach heights of over 1 km on Earth, and are thought to have been important during the Moon's volcanic history.

fluidized ejecta Impact ejecta that behaves as a fluid due to the presence of a *volatile* such as liquid water. The presence of liquid water in impact ejecta greatly enhances the mobility of the ejected debris, converting the dry fragmental ejecta flows characteristic of lunar craters to fluid debris flows similar to terrestrial mud flows. On Mars, this results in an unusual form of ejecta blanket associated with craters 5–15 km in diameter, with petal-like lobes that end in a low concentric ridge or outward facing escarpment. See also rampart craters.

fluvial processes Processes that are the result of water flow within (and at times beyond) a stream channel, bringing about the erosion, transfer and deposition of sediment.

fly-by A spacecraft on a trajectory that does not go into orbit around a planet. See also orbiter.

free-fall speed The maximum speed achieved by an object falling freely under gravity through an atmosphere.

gamma-rays Electromagnetic waves with the highest frequencies, above the highest frequencies of X-rays.

gas giant See giant planets.

genetic code The means by which genetic information is stored as sequences of nucleotides in DNA.

giant planets Jupiter, Saturn, Uranus and Neptune are the giant planets. They have radii greater than 24 000 km, densities close to 1×10^3 kg m^{-3}, are predominantly or wholly fluid, and are located in the outer part of the Solar System.

granite A light-coloured, coarse-grained igneous rock. It typically occurs as large bodies which are emplaced at depth within the Earth's continental crust.

habitable zone The region surrounding a star throughout which the surface temperatures of any planets present might be conducive to the origin and development of life. The temperatures required are generally taken to be those necessary for liquid water to exist on a planet's surface.

halophile An extremophile that thrives in extremely saline environments.

hot spots Areas with elevated levels of volcanic activity. On Earth hot spots can occur on plate margins, or within plates. The hot spot is thought to be stationary, or nearly so, and to produce volcanoes intermittently as the plate moves over it. It has been suggested that mantle plumes lie beneath hot spots.

hyperthermophile An extremophile that thrives in extremely high-temperature environments, up to about 105 °C, with a few tolerating 121 °C, and generally fails to multiply below 80 °C.

ice Sometimes used simply to refer to frozen water, this can also mean other *volatiles* in a frozen state (either singly or in a mixture) such as methane, ammonia, carbon monoxide, carbon dioxide and nitrogen. Frozen water is sometimes distinguished (for clarity) as water-ice.

igneous A term that refers to rocks formed by the cooling and crystallization of magma (ignis is Latin for 'fire').

impact cratering The process of formation of craters on the surfaces of solid planetary bodies through high-speed impacts. Impact structures may range in size from millimetres to thousands of kilometres.

impact craters See impact cratering.

infrared radiation Electromagnetic waves with frequencies or wavelengths between those of visible light and microwaves.

intrusive A body of rock, usually igneous, that is emplaced within pre-existing rocks.

kinetic energy The energy that an object possesses because of its motion.

Kuiper Belt The region of the Solar System, beyond the orbit of Neptune (30 AU) containing many icy planetesimals and cometary nuclei. The belt is thought to contain 10^7–10^9 bodies.

lander A spacecraft, or part of a spacecraft, designed to land on the surface of a planetary body.

lava see lava flow.

lava flow A mass of molten rock (or, in the icy satellites, a melt produced from ices) that is erupted onto the surface of a planetary body. Before it reaches the surface, this melt is more properly known as magma. After it has been erupted it is known as a lava flow. Once it has cooled down, the mass of solidified rock may still be described as lava or lava flow.

limestones Sedimentary rocks composed primarily of carbonate minerals.

lipids Organic compounds related to fats, also providing an energy store.

lithosphere The outer rigid shell of a planetary body, overlying the asthenosphere.

magma A hot melt of silicate composition containing dissolved volatiles, which forms the raw material for volcanism before eruption. Magma is not quite synonymous with lava, since lavas are often partially crystallized and have lost large amounts of gas. (Cryovolcanism could be said to involve watery magmas.)

mantle The compositionally distinct layer of a differentiated planetary body that overlies the core. It may be, as in the case of the Earth, overlain by a chemically distinct crust.

mantle plumes Localized, hot, buoyant material in the Earth's mantle which is hypothesized to originate near the mantle–core boundary.

marble A metamorphic rock composed largely of recrystallized limestones.

mesophiles Organisms that grow best at temperatures between 15 °C and 50 °C.

mesosphere A layer of the Earth's atmosphere lying between the stratosphere and thermosphere (50–87 km in altitude) in which temperature decreases with altitude.

metamorphic A term that refers to rocks whose texture or mineralogy has been changed through the action of heat and/or pressure.

microwaves Electromagnetic waves with frequencies or wavelengths between those of infrared radiation and radio waves.

mid-ocean ridge Term used to refer to the young, hot and therefore shallow part of the Earth's ocean floor occurring at a plate boundary where the plates are moving apart.

minor bodies Any of the many small rocky, icy or metallic objects in the Solar System, which are not classified as planets or their satellites.

molecules The smallest freely existing parts of a substance that retain the chemical identity of the substance. Molecules do not have to contain atoms of different elements.

monomers An individual organic unit which can be linked with similar units to form a polymer.

mudstones Extremely fine-grained sedimentary rocks formed from consolidated mud.

nucleic acids There are two types of nucleic acid: DNA (deoxyribonucleic acid) and RNA (ribonucleic acid). DNA and RNA are polymers in which the monomer is called a nucleotide. DNA and some types of RNA are chemicals that carry the genetic information of cells.

nucleotides Monomers that form a nucleic acid. Each monomer consists of a phosphate group and a base attached to a ribose molecules or deoxyribose molecule.

Oort cloud A spherical cloud of possibly 10^{12} comets (with a total mass about 25 times that of the Earth) that is hypothesized to extend out to tens of thousands of AU from the Sun.

orbiter A spacecraft designed to go into orbit around a planetary body. See also lander.

organic In chemistry denotes compounds containing carbon.

ozone A gas with molecules that each consist of three atoms of oxygen.

partial melting Rocks do not have a fixed melting temperature, like pure ice, but melt over a range of temperatures. Partial melting in the low part of the range liberates some components preferentially, leading to liquids (magmas) different in composition from the initial rock. This is an important process in planetary differentiation.

peridotite The rock type of which the uppermost part of the Earth's mantle is composed.

photosynthesis The synthesis, by green plants (including algae) and some bacteria, of organic compounds using the energy of sunlight. Carbon dioxide is converted to carbohydrate through addition of hydrogen atoms derived (in all photosynthetic organisms except certain bacteria) from water, and oxygen is produced as a consequence.

piezophiles Extremophiles that are able to tolerate high pressures.

planetary embryo A hypothetical body of something like one-hundredth to one-tenth the mass of a planet, produced as the end-product of runaway growth of planetesimals.

planetesimals A body roughly 100 m to 10 km across, formed by coagulation of dust grains in the solar nebula, or a somewhat larger body produced by accretion of smaller planetesimals. Accretion of smaller planetesimals onto larger ones may have occurred in a runaway fashion, leading to the production of planetary embryos.

plate recycling Outward transfer of heat in a planetary body through the creation of new, hot, lithosphere at the edges of tectonic plates, and the removal (subduction) of old, cold, lithosphere down into the asthenosphere. Thanks to plate tectonics, this is the main means of lithospheric heat transfer in the Earth.

plate tectonics A description of the behaviour of the Earth's lithosphere, which is broken into seven major plates and a few minor ones. These plates can be thought of as rigid, and they can move about because of the weakness of the underlying asthenosphere. Plates are added to at spreading plate boundaries and destroyed at a globally averaged equivalent rate at subduction zones.

polymers Large molecules in which a group of individual organic units are repeated.

proteins Large organic compounds made of chains of amino acids.

protoplanetary disc A disc of gas and dust around a star from which planets may form (see also solar nebula).

psychrophiles Extremophiles able to tolerate temperatures of less than 15 °C.

pyroclastic materials Fragmentary ('fire-broken') volcanic rocks, erupted as solid particles rather than molten lava.

radio waves Electromagnetic waves with the lowest frequencies/longest wavelengths, extending from the lowest frequency/longest wavelength microwaves.

rampart craters Some impact craters on Mars, 5–15 km in diameter, exhibit striking aprons of continuous ejecta, extending about 1 crater diameter, resembling flower petals, which terminate in elevated ridges or 'ramparts'. They are thought to indicate that the target material contained permafrost ice.

resolution The ability of an optical instrument to distinguish fine detail.

respiration The metabolic process through which biological material initially formed by photosynthesis is converted to carbon dioxide and water, liberating useful energy.

RNA (ribonucleic acid) A polymer in which the monomer is a composite molecule consisting of a phosphate group joined to a ribose molecule.

sandstones A medium- to coarse-grained sedimentary rock composed largely of the mineral quartz.

saturated In cratering terms, a surface is said to be saturated when a new impact crater can only be formed by overprinting and thus obliterating older ones.

secondary craters Clumps of ejecta from major impacts are often large enough to form secondary craters when they fall back. Secondary craters up to a few kilometres in diameter are so common that they complicate planetary cratering statistics.

sedimentary rocks A term applied to rocks that have formed from sediment deposited by water or wind.

seismic waves Vibrations transmitted through a planetary body.

shield volcano Broad volcano with gently sloping side.

silicates Minerals containing silicon and oxygen, and usually metallic elements. The term silicates may be used to refer to material (such as most rocks) formed principally of silicate minerals.

slate A metamorphic rock produced by the effect of pressure on a fine-grained sedimentary rock.

solar nebula The hypothetical cloud of gas and dust within which the Sun and the other constituents of the Solar System formed, according to the nebula theory. The solar nebula gave rise to a protoplanetary disc as the nebula flattened into a disc.

solid-state convection Convection occurring in a solid, without the necessity of melting, at high pressures and temperatures.

spatter cones Small volcanic cones (usually 5–20 m high) produced by material thrown out as spatter.

spatter Fluid basaltic pyroclastic material which accumulates by fallout from a volcanic eruption column to form a rampart around the vent.

star A self-luminous celestial body consisting of a mass of gas held together by its own gravity that, at some stage of its life, produces energy by nuclear reactions, mainly the conversion of hydrogen into helium.

stratum (plural strata) A layer of rock.

stratosphere A layer of the Earth's atmosphere at 20–50 km altitude, in which the temperature increases with height.

subduction zones See plate tectonics.

terrestrial planets Mercury, Venus, Earth and Mars are the terrestrial planets. They are dominantly rocky objects, with iron-rich cores and silicate mantles, and densities of $3.9–5.5 \times 10^3 \, kg \, m^{-3}$.

thermophiles An extremophile that thrives in environments where the temperature is high, typically up to 80 °C.

thermosphere The outermost region of an atmosphere, characterized by increasing temperature with altitude.

tidal heating Heating of a planetary body caused when varying tidal forces continually distort it.

transient cavity When a high-speed impact takes place, the material at the target site is highly compressed and depressed in a transient cavity. The final shape of the crater depends on the ejection of material from this cavity, and subsequent slumping and settling of material from its walls.

troposphere The lower region of a planetary atmosphere in which mixing occurs by convection.

ultraviolet (UV) radiation Electromagnetic waves with frequencies or wavelengths between those of X-rays and visible light.

visible light Electromagnetic waves with frequencies or wavelengths between those of ultraviolet radiation and infrared radiation. Our eyes are sensitive to visible light.

volatiles Elements or compounds that melt or boil at relatively low temperatures, or (equivalently) condense from a gas at a low temperature. Hydrogen, helium, carbon dioxide, and water are examples.

volcanic bombs A relatively large (>64 mm) lump of rock thrown out during an explosive eruption.

water-ice Frozen water is sometimes distinguished (for clarity) as water-ice. Ice can also mean other *volatiles* in a frozen state (either singly or in a mixture) such as methane, ammonia, carbon monoxide, carbon dioxide and nitrogen.

weathering Breakdown and disintegration of rocks and minerals on a planet's surface by physical and chemical processes.

xenoliths An inclusion of a pre-existing rock preserved within a later igneous (e.g. volcanic) rock.

xerophiles Organisms that can tolerate extreme desiccation by entering a state of apparent suspended animation, characterized by little water within their cells and a cessation of biological activity.

X-rays Electromagnetic waves with frequencies or wavelengths between those of gamma-rays and ultraviolet radiation.

ACKNOWLEDGEMENTS

Grateful acknowledgement is made to the following sources for permission to reproduce material within this book.

Cover photo and title page: NASA.

Figures

Figure 0.3a–c Mary Evans Picture Library; *Figure 0.4* Marc W. Buie/Lowell Observatory.

Figures 1.1, 1.4, 1.5, 1.6, 1.7, 1.8, 1.9, 1.10, 1.11, 1.12, 1.14, 1.15, 1.16, 1.17, 1.19, 1.20, 1.22, 1.23, 1.26, 1.27 and 1.28 NASA; *Figure 1.13* USGS; *Figure 1.18* Russian Academy of Sciences/RNII KP/IPPI/TsDKS; *Figures 1.21, 1.24 and 1.25* NASA/JPL/Malin Space Science Systems; *Figures 2.1, 2.2, 2.3, 2.4, 2.5, 2.6 and 2.10* NASA.

Figure 2.7 Steward Astronomy, University of Arizona. *Figure 2.9* ESA.

Figures 3.1, 3.2 and 3.9 NASA/JPL/Malin Space Science Systems; *Figures 3.3 and 3.6* NASA; *Figure 3.8* JPL.

Figures 4.4, 4.5, 4.6, 4.7, 4.8, 4.9, 4.10, 4.16, 4.18, 4.20, 4.21, 4.22, 4.23, 4.24, 4.25, 4.27, 4.28, 4.31, 4.32, 4.33, 4.35, 4.36, 4.37, 4.38, 4.39, 4.40, 4.41, 4.42, 4.44, 4.45, 4.46 and 4.47 NASA; *Figures 4.11, 4.13 and 4.14* Steve Self/Open University *Figures 4.12, 4.15 and 4.19* USGS; *Figure 4.26* NASA/JPL/Arizona State University; *Figure 4.29* Institute of Geological Sciences; *Figure 4.30* © Lunar and Planetary Institute; *Figures 4.34b and 4.43* NASA/JPL/Malin Space Sciences Systems.

Figure 5.1 Robert Thom; *Figure 5.2* The Natural History Museum, London; *Figures 5.4, 5.9, 5.12, 5.16, 5.17, 5.18, 5.24, 5.29, 5.30, 5.31, 5.36, 5.37, 5.40b and 5.40c* NASA; *Figure 5.7* I. D. J. Burdett; Figure 5.8 Simon Conway Morris, University of Cambridge; *Figure 5.14* NASA/US Geological Survey; *Figure 5.23* NASA/Arizona State University; *Figures 5.25, 5.26, 5.27 and 5.28* NASA/JPL/Malin Space Sciences Systems; *Figure 5.33* NASA/JPL/University of Arizona/Los Alamos National Laboratories; *Figure 5.34* NASA/JPL; *Figure 5.40a* Copyright © Proszynski I S-ka SA 1999–2001. Wszystkie prawa zastrzezone; *Figure 5.41* Everett Gibson (NASA/JSC).

Every effort has been made to contact copyright holders. If any have been inadvertently overlooked the publishers will be pleased to make the necessary arrangements at the first opportunity.

INDEX

Entries and page numbers refer to key words that are printed in **bold** in the text.

A

accretion 46
acidophiles 117
advection 60
Aeolian processes 94
alkaliphiles 117
asteroid belt 23
asthenosphere 59
astronomical unit 4
atmospheric structure 127

B

basalt 39
breccias 39
brecciated 83

C

calderas 73
carbohydrates 108
carbon fixation 137
carbonates 39
core 50
crust 50
cryovolcanism 8
cytosol 111

D

Darwinian evolution 104
differentiation 48
DNA 109
double helix 110

E

ecliptic plane 4
effusive volcanism 68
electromagnetic radiation 33
electromagnetic spectrum 33
endoliths 119
enzymes 109
erosion 38
eruption columns 71
explosive volcanism 68
extremophiles 113
extrusive 38

F

fire fountains 70
fluidized ejecta 91
fluvial processes 94
fly-by 20
free-fall speed 81

G

gamma-rays 33
gas giant 7
genetic code n111
giant planets 5
granite 39

H

habitable zone 106
halophile 118
hot spots 61
hyperthermophile 114

I

igneous rocks 38
impact cratering 80
impact craters 8
infrared radiation 33
intrusive 38

K

kinetic energy 83
Kuiper Belt 13

L

lander 20
lava 15
lava flow 69
limestones 39
lipids 108
lithosphere 59

M

magma 38
mantle 50
mantle plumes 61
marble 40
mesophiles 114
mesosphere 127
metamorphic rocks 38
microwaves 33
mid-ocean ridge 61
minor bodies 5
molecules 105
monomers 109
mudstones 39

N

nucleic acids 108
nucleotides 109

O

Oort cloud 5
orbiter 20
organic 105
ozone 115

P

partial melting 48
peridotite 52
photosynthesis 136
piezophiles 117
planetary embryo 47
planetesimals 47
plate recycling 61
plate tectonics 52
polymers 109
proteins 108
protoplanetary disc 46
psychrophiles 114
pyroclastic materials 69

R

radio waves 33
rampart craters 91
resolution 30
respiration 137
RNA 111

S

sandstones 39
saturated 93
secondary craters 93
sedimentary rocks 38
seismic waves 51
shield volcano 73
silicates 37
slate 40
solar nebula 46
solid-state convection 59
spatter cones 72
star 3
stratum (plural strata) 38
stratosphere 127
subduction zones 61

T

terrestrial planets 5
thermophiles 115
thermosphere 127
tidal heating 58
transient cavity 83
troposphere 127

U

ultraviolet (UV) radiation 33

V

visible light 33
volatiles 47
volcanic bombs 72

W

weathering 38

X

xenoliths 52
xerophiles 116
X-rays 33